COMPASS AND RULE

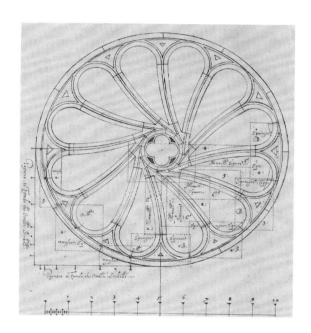

Compass and Rule

Architecture as Mathematical Practice in England
1500–1750

Anthony Gerbino and Stephen Johnston
with a contribution by Gordon Higgott

Yale University Press, New Haven and London
in association with
Museum of the History of Science, Oxford, and Yale Center for British Art

This publication accompanies the exhibition
Compass and Rule: Architecture as Mathematical Practice in England, 1500–1750
organized by the Museum of the History of Science, Oxford
at the Museum of the History of Science from 16 June to 6 September 2009
and the Yale Center for British Art, New Haven, from 18 February to 30 May 2010

Copyright © 2009 by the Museum of the History of Science, Oxford

All rights reserved. This book may not be reproduced, in whole or in part, including illustrations, in any form (beyond that copying permitted by Sections 107 and 108 of the U.S. Copyright Law and except by reviewers for the public press), without written permission from the publishers.

Designed by Sally Salvesen
Printed in Italy by Conti Tipocolor S.p.A., Florence

Library of Congress Cataloging-in-Publication Data
Gerbino, Anthony.
 Compass and rule : architecture as mathematical practice in early modern
England, 1500-1750 / Anthony Gerbino and Stephen Johnston.
     p. cm.
 Includes bibliographical references and index.
 ISBN 978-0-300-15093-3 (alk. paper)
 1. Architectural practice—England—History. 2. Architecture—England—History. 3. Architects—England.
4. Mathematics—England—History. 5. Mathematicians—England. 6. Architecture and mathematics—England—
History. 7. Professions—Social aspects—England—History.
I. Johnston, Stephen, 1961– II. Title. III.
Title: Architecture as mathematical practice in early modern England, 1500-1750.
 NA1995.G47 2009
 720.942'0903--dc22
                    2009008861

Illustrations:
Half title: detail from figure 48
Frontispiece: detail from figure III

Contents

	Foreword	7
	Acknowledgements	8
	Lenders to the Exhibition	9
	Introduction	11

PART I

1	Medieval Drawing and the Gothic Tradition	17
2	The Paper Revolution: The Origin of Large-scale Technical Drawing under Henry VIII	31
3	The Mathematical Practitioner and the Elizabethan Architect	45
4	The Vitruvian Model: Inigo Jones and the Culture of the Book	65
5	Vision, Modelling, Drawing: Christopher Wren's Early Career	83
6	Structure and Scale: The Office of Works at St Paul's	97
7	Gentlemen, Practitioners, and Instrumental Architecture	111
8	Raised High, Brought Low: Architecture and Mathematics around George III	131

PART II

Geometry and Structure in the Dome of St Paul's Cathedral — 155
by Gordon Higgott

PART III

Catalogue Checklist — 173

Notes	189
Bibliography	199
Index	205
Photographic Acknowledgements	208

Foreword

Partnership seems to lie at the heart of every aspect of *Compass and Rule*. The research, writing, and curatorial planning have been collaborations between two scholars, Anthony Gerbino and Stephen Johnston. They bring together two academic disciplines, the histories of architecture and of science, whose modern separation has impoverished our understanding of the period covered by this project. Historians of science will be excited to see architecture provide a rich ground for working through some recent revisions in their approach to early-modern geometrical practice, extending their understanding of the broad range of the geometrical arts. Historians of architecture will find their recent determination to study process as much as product strengthened through a similar adjustment that has been made in the history of science.

Partnerships with lenders have been vital to this project and we are grateful to those institutions and individuals who have offered precious works with the greatest generosity and enthusiasm. We would also like to express our appreciation to Gordon Higgott for his frequent help and advice as well as his contribution to the catalogue. The book is a model of interdisciplinary scholarship and will serve as a valuable permanent record of the many creative collaborations that have made this project possible, but an exhibition is the most appropriate and complete medium for a presentation of process. The instruments, drawings, and books in their immediate material presence offer the most engaging and compelling experience of this historical narrative, and even in the catalogue the authors have chosen to present their account through the succession of objects experienced in the galleries.

The partnership between the two organizing institutions – the Museum of the History of Science at the University of Oxford and the Yale Center for British Art in New Haven – has reinforced the interdisciplinary nature of the project. Through this collaboration, two distinct groups of museum goers who too rarely find shared interests will be brought together on common ground. Although science and art have generally been considered as distinct realms of intellectual pursuit, in recent years a number of projects have attempted to combine them, some more successfully than others; how much more natural to centre one in the early-modern period, when these endeavours were, in fact, inseparable.

Jim Bennett
Director
Museum of the History of Science

Amy Meyers
Director
Yale Center for British Art

Acknowledgements

The authors are particularly grateful to our lenders and to the many staff who have provided access to, and expertise on, collections, archives, and library materials. Without their help the catalogue research and the exhibition would have been impossible. In addition to our institutional lenders, Howard Dawes has been exceptionally generous in his contribution. We are also extremely pleased that the Yale Center for British Art has agreed to host the exhibition as a second venue and are deeply gratified by their encouragement, enthusiasm, and support. We would also like to thank the following individuals for aid, advice, and feedback: Derek Adlam, Philip Beely, Martin Biddle, Mary Bosworth, Filippo Camerota, James Campbell, Isabelle Carré, Diane Clements, David and Yola Coffeen, Cliff Davies, Anthony Geraghty, Jeroen Goudeau, Jacques Heyman, Hester Higton, Maurice Howard, Edward Impey, Michael Korey, Alex Marr, Robert Martensen, Richard Morris, Pascal Mychalysin and the masons at Gloucester Cathedral, Richard Ovenden, Joanna Parker, Stephen Porter and James Thompson of the London Charterhouse, Frank Salmon, Jonathan Smith of Trinity College, Cambridge, Matthew Walker, Steve Walton, Anthony Beckles Willson, Michael Wright, David Yeomans, and, finally, our editor, Sally Salvesen, for her infinite patience and understanding. Their assistance is warmly acknowledged. Jim Bennett and the staff at the Museum of the History of Science have provided essential support throughout the project. We acknowledge, finally, the Arts and Humanities Research Council, whose generous research grant made the project possible.

Lenders to the Exhibition

All Souls College, Oxford
The Bodleian Library, University of Oxford
The British Library
The British Museum
Corpus Christi College, Cambridge
Howard Dawes
Emmanuel College, Cambridge
Gainsborough's House Society
Jesus College, Cambridge
King's College, London
London Metropolitan Archives
The Museum of Archaeology and Anthropology, University of Cambridge
The National Archives
National Museums Scotland
The Royal Collection
Royal Institute of British Architects
The Dean and Chapter of St Paul's Cathedral
Science Museum, London
Sir John Soane's Museum
The Wellcome Library
Whipple Museum of the History of Science, University of Cambridge
Winchester College
Worcester College, Oxford

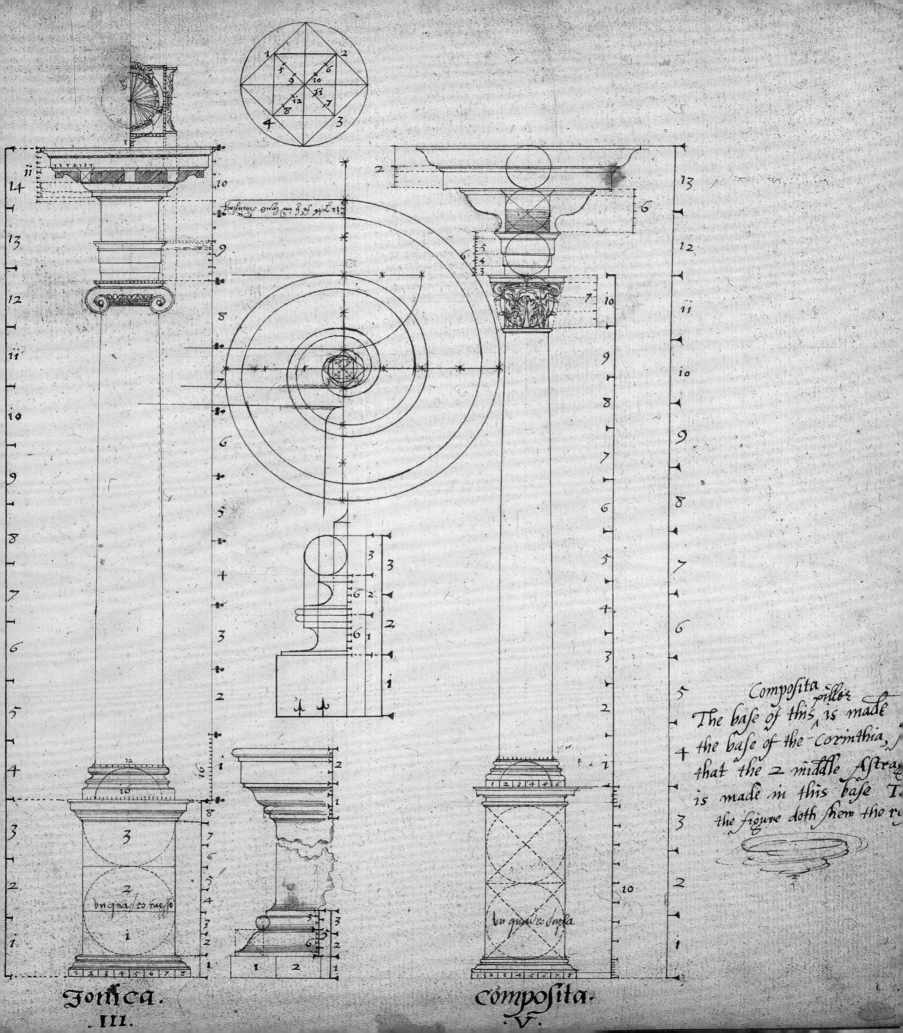

Ionica.
III.

Composita.
V.

Composita pillor
The base of this is made
the base of the Corinthia,
that the 2 middle Astra
is made in this base
the figure doth shew the r

Introduction

It is generally accepted that the introduction of classical architecture in England coincided with a new public identity for building practitioners. With the spread of Renaissance culture, the task of the architect – as opposed to the simple builder – was increasingly equated with the process of design as a stage separate and superior to that of actual construction. In this view, design was seen as a primarily intellectual process, which prioritized the inventive and creative faculties of the designer over and above the putatively menial procedures of actual building. This shift brought about a number of changes that were essential to the evolution of the profession. It allowed the architect a greater social standing relative to other artisans and to separate himself professionally from the corporate organization of masons. More importantly, it allowed for a closer and more intimate working relationship with the aristocratic patron. Inevitably, this process was neither rapid nor uniform, especially in England, where the tradition of the artisan-builder was far more tenacious than elsewhere and where the classical style was often resisted as a foreign import. Nevertheless, by the mid-eighteenth century, the transformation was largely complete.[1]

Compass and Rule addresses a largely unexplored dimension of this story: the extent to which this professional identity was based on a growing expertise in and identification with the mathematical arts and sciences. Mathematics provided a natural paradigm for early modern architects. First, the art was almost wholly dependent on geometrical or arithmetic operations of some form or another. The process of design itself – insofar as it involved the application of proportional or compass-based rules – was largely defined by them, as were many other basic tasks. Surveying, cost estimates, bookkeeping, and even the use of routine graphic techniques all entailed a certain amount of mathematical training. Moreover, as practitioners of an art increasingly identified with a superior realm of 'theory', early modern architects were receptive to advances in contemporary mathematical sciences in a way that was not true of their predecessors. This work seeks to explore issues and questions raised by that situation. To what extent could the architect be considered a 'mathematical practitioner'? What role did mathematics play in architectural practice and in building technology? How were cultural references to mathematics deployed by building practitioners and received by the wider public?

Our work is driven by two principal themes. The first explores the changing nature and role of 'design', conceived as a discrete feature of architectural practice, while the second tracks the way in which this new concept was used as a support for social and professional legitimacy. In both cases, mathematics – geometry in particular – provided building practitioners with an important analogy and resource for their work. As an operative technique involving compass and ruler, geometry helped to define the manual, instrumental character of the design process. As an 'applied' technology, it linked architecture to the wider constellation of mathematical disciplines. Finally, as a 'science', it served to invest design with the intellectual and historical authority of an ancient tradition. Over time, the reiterated bond between architecture and mathematics would help form the basis of a new professional identity.

The emphasis on mathematical practice somewhat alters our view of early modern architecture, for it concerns less the visual character of the designed building than the *process* of design itself and, in particular, the physical and conceptual tools that the architect brings to it. Such a prism also modifies our understanding of the architect's material culture, and our study reflects this unusual perspective in two important ways. First, instruments are foregrounded. From simple builders' tools, such as the square, level, and compass, to more sophisticated aids for calculating and measuring, these objects were the means by which mathematics was translated into practice, both on paper and at the work site. Drawings, too, are dominant, but they have been chosen and are discussed from a distinct standpoint, one that concentrates on the origin and development of specific representational conventions. In other words, we explore the expanding role of the medium itself, both as a motor of the design process and as a mark of the draughtsman's technical expertise. In England, for example, scaled paper plans did not become a normative design tool, either for conceptualizing or executing building projects, until the second half of the sixteenth century. The very success of the technique has served to obscure its earlier history.[2]

The term 'mathematical practice' is multivalent, and it may be worthwhile to set out its historiographical background. It originated in the work of E. G. R. Taylor as a way of identifying an English didactic tradition of mathematical arts and sciences. In Taylor's telling, the movement to improve English mathematics took shape in the mid-sixteenth century, as the combined work of navigators, shipwrights, instrument makers, and gentlemen mathematicians. A heterogeneous assembly, they were united by a vision of mathematics as a practical and worldly activity, which would serve to advance diverse crafts and disciplines via their common foundations in geometry and arithmetic. The 'mathematical practitioners' of Tudor and Stuart England emphasized the importance of instruments for observation, measurement, and calculation, and the value of maps, charts, and plans produced by those selfsame instruments. The programme was at once civic-minded and inner-directed, intended to bring both 'useful' advantages to the commonwealth and intellectual satisfaction to its adherents. This new mathematical culture was articulated as an indigenous, vernacular tradition and expressly promoted through the medium of the printed book.[3]

Although this movement embraced architecture, it was distinct from the Italian–Vitruvian conception of the art. As Rudolf Wittkower observed many years ago, the mathematical practitioners did not seek to provide a complete or systematic doctrine of design or construction. In fact, they hardly addressed the aesthetic or functional requirements of building at all, but were instead concerned with discrete, practical problems of quantity and land surveying, and in the description of new instruments. Their influence is particularly evident in surveyors' manuals, for which mathematical practitioners supplied a number of standard texts. Leonard Digges's *A Boke named Tectonicon* of 1556, for example, has often been singled out as an inaugural work of both land and material surveying.[4] The dual focus reflects a peculiarity of architectural practice in early modern England, which was often joined to a range of administrative responsibilities related to the care of estates.[5]

More recently, historians of science have deployed the concept of mathematical practice not merely as a descriptive notion – a way of classifying individuals and their activities – but as an explanatory one. This body of research has sought to give a more prominent role to the practical mathematical sciences in the origins of the 'scientific revolution', particularly in the emergence of the 'mechanical' natural philosophy of the later seventeenth century. As Jim Bennett and others have argued, many of the investigative techniques that transformed the practice of natural philosophy can be traced to the mathematical tradition. These included the use of instruments to measure and quantify, the emphasis on repeated trials and 'experience', and the use of explanatory models and drawings. The disciplines of mathematics and natural philosophy had traditionally been kept separate: the latter was concerned with the causal and theoretical understanding of nature, while the mathematical sciences were oriented toward the development of practical techniques tending toward useful ends. It was the interaction between the two domains – manifest in the adoption of mathematical and mechanical explanations for natural phenomena – that came to characterize modern scientific practice.[6]

This argument entails a second claim, more closely related to the themes of this work: that to preserve an authentic, 'actor's' view of early modern mathematics, historians must avoid making rigid distinctions between 'high' and 'low' – or theoretical and practical – traditions of the discipline. The concerns of surveyors, engineers, and navigators are seen, in this reading, to overlap with those of mathematicians. Many practitioners were interested in questions of higher mathematics and natural philosophy as they pertained to the advancement of their art, while mathematicians frequently looked to problems and phenomena thrown up by craft practice. The quest for longitude and the problem of projectile motion are perhaps the most famous instances of this combined effort, but the same dynamic existed across a number of different fields. The worlds of the 'scholar' and 'craftsman' – those hoary categories of mid-twentieth-century scholarship – are now justly seen not as mutually exclusive but as mutually reinforcing.

This body of research helps to explain the unusual careers of some late seventeenth-century *virtuosi*, Christopher Wren's in particular. His scientific interests can easily strike a modern observer as heterogeneous and scattershot, and it has often been difficult for historians to account for the way they cohered, a problem made more acute by his marked concern for useful inventions and 'low'-status forms of practice. The preponderant role that architecture came to play in his career has proved especially perplexing. Why would such an accomplished mathematician and natural philosopher abandon these pursuits for design and building? Did he see the two parts of his career as separate and parallel – much as they have been defined by modern historians – or were they more integrally combined than first appears? Jim Bennett's study of Wren, which remains the standard treatment of his scientific work, attempts to answer these questions by placing him within the historical context of English mathematical practice.[7] Our goal is to show that architecture, too, had a 'mathematical' history that allowed Wren and others to incorporate it into a larger notion of the 'mathematical sciences'. As we hope to show, his particular combination of interests – often ascribed uncritically to his unique and powerful genius – was, in fact, part of a broader historical phenomenon, with roots in the very origins of Renaissance architecture in Britain.

PART I

Chapter One
Medieval Drawing and the Gothic Tradition

The link between architecture and the 'high' or theoretical tradition of geometry is traditionally seen as a creation of Renaissance Neoplatonism, in tandem with the 'rediscovery' of Vitruvius and the emergence of the Italian architectural treatise. This association was created, in part, by Renaissance architects themselves. In the opening pages of the *Primo Libro* (1545), for example, Sebastiano Serlio complained of those 'who today bear the title, "architect" but who do not know how to give a definition of a point, a line, a plane or body, or say what correspondence and harmony are'.[1] For Serlio, Euclidian geometry provided an elevated standard of knowledge, appropriate to a new class of building practitioners and to a new *all'antica* style. The charge would become a commonplace of architectural literature throughout the early modern period.

This implication – and its attendant chronology – should, however, be qualified. In England, appeals on the part of builders to the authority of mathematics long predate the introduction of classical architecture into the country. Indeed, the earliest surviving masons' ordinances provide evidence of such a linkage as early as the late fourteenth or early fifteenth century. The famous Cooke and Regius manuscripts, now in the British Library, equate the art of masonry with the 'science of geometry', characterizing it, moreover, as the origin and 'first cause' of the seven liberal arts. The 'worthy clerk Euclid' plays a founding role in both texts, each of which provides a long legendary history of the craft, from its invention in biblical times to its introduction in England under the Anglo-Saxon king Athelstan. Most importantly, the texts distinguish between the 'practice of the science' and its 'speculative' part, implicitly separating the conventions of proportion, composition, and planning from the physical task of stonecutting itself.

As portraits of craft experience and examples of artisanal expression, the texts are unique. While probably written with the assistance of a clerk, they are virtually alone among the extant sources for medieval building crafts in that they are about and for masons themselves. In this context, it is not surprising that the conception of geometry that they present is emphatically not scholastic. Indeed, the rather forced emphases on the liberal arts tradition and on the distinction between 'theory' and 'practice' are clearly intended to ennoble a popular understanding of geometry that was essentially physical, practical, and instrumental. As the author of Cooke asserts, the primacy of this science depends on its importance for human life: 'for there is no art or handicraft wrought by man's hands that is not wrought by Geometry' (102–105). The mythical Euclid of the manuscripts plays an analogous role in the text, one far removed from the Greek author of the *Elements*. This 'Euclid' is a surveyor. Having taught the first masons their craft, the Cooke manuscript claims, he gave it the name of geometry 'on account of the parcelling out of the ground which he had taught the people at the time of making the walls and ditches ... to keep out the water' (510–16). The legend draws on Isidore of Seville's etymology of the word

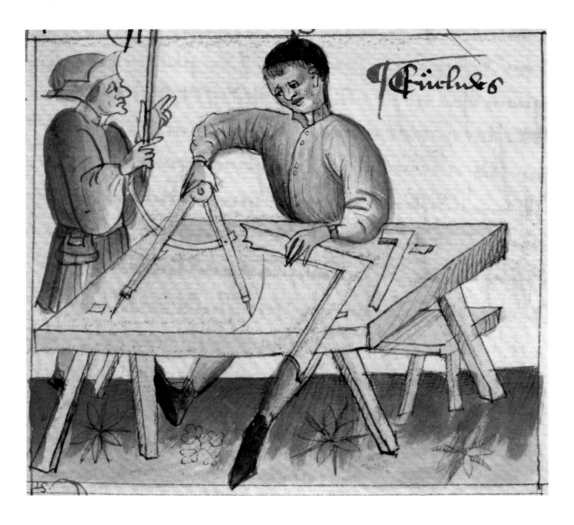

4. Euclid as medieval mason, fifteenth century. British Library, Additional MS 15692, f. 29v

'geometry' to elevate a mundane practical art.[2]

This definition of geometry was not idiosyncratic. John Lydgate, a rough contemporary of the Cooke and Regius authors, also identified Euclid with the mason's art and, in particular, his tools: 'By crafft of Ewclyde mason doth his cure, to suwe heos mooldes ruyle, and his plumblyne'.[3] The mathematician Robert Recorde encountered the same insistence in his dealings with craftsmen a century later: 'Carpenters, Karuers, Joyners, and Masons, doe willingly acknowledge that they can worke nothyng without reason of Geometrie, in so muche that they chalenge me as a peculiare science for them'.[4] The idea that geometry was a 'peculiar' science – private and distinctive to the building crafts – was apparently widespread. Of course, the mason's geometry had very little to do with the historical Euclid; in this respect, Serlio was right. The image was, nevertheless, characteristic of the kind that builders would continue to invoke over the next few centuries. Although Serlio claimed to represent a more accurate and scholarly understanding of Greek mathematics, the shift in attitude that he represented was one of degree, not of kind. The identification of geometry with the process of composition, planning, and setting-out was one that architects would continue to deploy throughout the early modern period.

The 'geometry' of the Cooke and Regius manuscripts almost certainly refers to the Gothic tradition of architectural drawing, at once the most geometrical component of medieval building and the most likely precursor to the Renaissance notion of design. Unlike their present-day counterparts, medieval architectural drawings consisted mostly of full-scale details used for shaping and assembling stones. Many fragments survive, incised onto walls or plaster tracing floors specifically set aside for the purpose. The well-known tracing floors at York Minster and Wells Cathedral, for example, still retain identifiable outlines of window tracery and vaulting ribs. Robert Branner has argued that the technique arose during the thirteenth century in response to the complex stereotomy of Gothic buildings, because it allowed masons accurately to work out the cutting of stones in detail beforehand. Full-scale drawing was also used for making templates. These smaller designs could be set out on drawing tables, before being transferred to wood or metal moulds (figure 4). Both forms of the technique remained a standard part of masonry design and construction for centuries.[5]

Medieval Drawing and the Gothic Tradition

Our study considers several remnants of this culture, including two mason's drawings. Although neither was made to full scale, they represent in different ways how drawing functioned as part of the stonecutting craft. The first is a fragment of a window design from the thirteenth-century chapel of St John's College, Cambridge, almost certainly depicting its original east window, which had been replaced during remodelling in 1511–16. The drawing is inscribed onto a stone slab, which was then used as building material and subsequently revealed during the chapel's demolition in 1869 (figure 5, catalogue 1). The object is evocative on several levels. First, it brings home the unity of design and construction in medieval practice: the drawing was clearly made not in a separate loft or tracing room but on site, probably in view of the wall in which the window was to be set. In such a context, the suitability of drawing surfaces was determined simply by what was closest to hand. Although drawings of this sort could also be made on parchment or paper, such materials were expensive. Stone could serve as easily, especially for project drawings made on site. This example was evidently made to work out the window's basic design features and to provide a reference for the masons charged with executing the full-scale version.[6]

The drawing is particularly useful because it bears evidence of the way it was set out. In reconstructing that process, we can understand what medieval masons may have meant when they referred to their art as a province of geometry. As Lon Shelby has shown, the term referred not to the deductive and demonstrative science but rather to a design process characterized by

5. Incised window sketch from the chapel of the Hospital of St John the Evangelist, late thirteenth century. University of Cambridge Museum of Archaeology and Anthropology. Catalogue 1

6. Setting-out of St John's Chapel window sketch (line drawing by Martin Biddle)

Compass and Rule

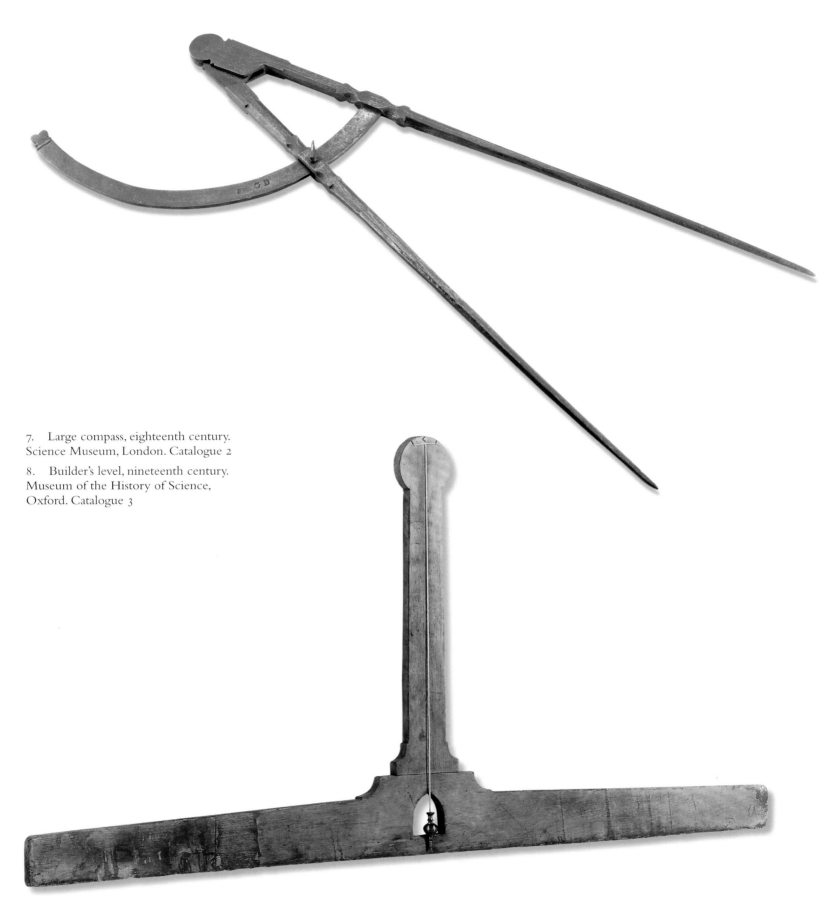

7. Large compass, eighteenth century. Science Museum, London. Catalogue 2

8. Builder's level, nineteenth century. Museum of the History of Science, Oxford. Catalogue 3

the rule-of-thumb manipulation of compass and ruler. In this case, the mason began by creating a baseline at the springing of the main arch, then dividing this line into units of three and eight. The compass marks are still apparent where he struck off the intervals (figure 6). These measurements gave him the width of the three subsidiary arches and that of the six window lights. Most of the other design features – including the three large circles in the head of the window and the three small circles under the intermediate arches – were derived from these basic proportions and the initial curves set out from them. What makes the drawing noteworthy, particularly in light of this somewhat rule-bound process, is its impromptu character. Stone is not the medium that we associate with sketching. The mason, however, had internalized the technique of constructive geometry to such a degree that he could freely score the slab with the metal point of his compass as though it were a pencil on paper.[7]

Expanding the concept of drawing in this way might lead us to think differently about the tools of the mason's trade. The great compass, set-square, and level are recurrent features on portrait busts and tomb slabs of master masons – as they are in painted illuminations of the medieval building site. The tools were emblematic of the craft. They had, however, a double significance, for they were not only tools of construction but also of constructive geometry, that is, drawing. It was in this sense that John Lydgate, quoted above, could associate Euclid with 'moulds, rule, and plumbline'. Examples of such tools can be found in the collections of the London Science Museum and the Museum of the History of Science (figures 7, 8, catalogue 2, 3). Typically later in date, these artefacts emphasize the continuity of the practice of full-scale drawing. Even as Gothic architecture gave way to Renaissance classicism in many parts of England, the tradition of full-scale drawing lived on, particularly for setting out and composing details. The template designs contained in the late sixteenth- and early seventeenth-century album of John Thorpe represent the persistence of this tradition well into the early modern period.[8]

The second drawing showcased in this section is a sketch contained in the fifteenth-century *Itineraries* of the traveller and antiquary, William Worcestre (or Worcester). This Bristol man had served for many years as secretary to Sir John Fastolf, a soldier and adventurer with extensive lands in Norwich. Worcestre, curious and

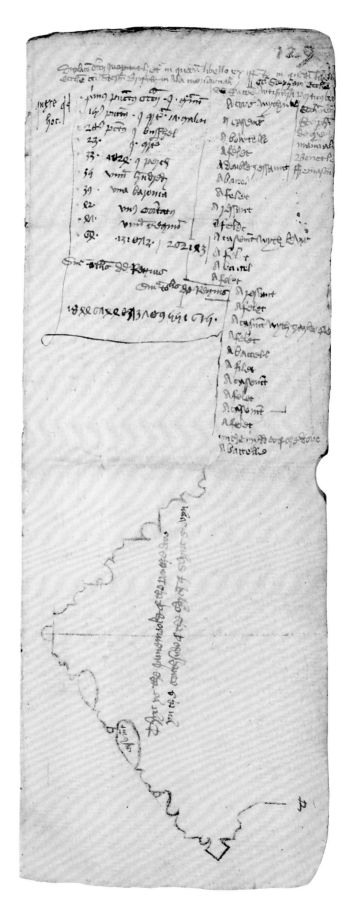

9. William Worcestre (and Benedict Crosse?), drawing of the south portal jamb of St Stephen's Church, Bristol, 1480. Corpus Christi College, Cambridge. Catalogue 4

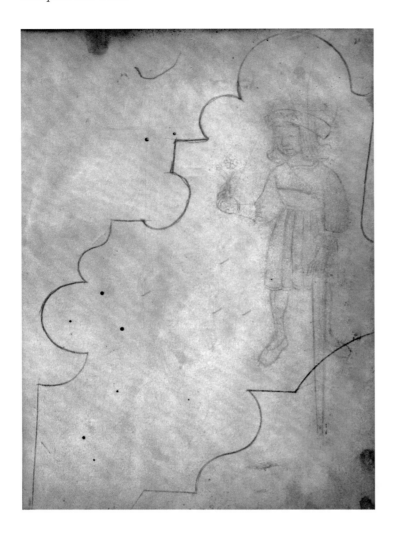

10. Tracing of a moulding profile in a medieval pattern book. Magdalene College, Cambridge, Pepys Library 1916, f. 21a

11. Plan of the court and kitchen of Winchester College, c.1394. Winchester College. Catalogue 5

with wide scholarly interests, was also an indefatigable compiler. In the latter part of his life, he devoted himself to travelling and, during journeys that he made from 1478–80, kept a notebook in which he recorded extensive observations and notes. These ranged widely, from the names of plants and animals to records of local saints and festivals to obituary lists, calendars, and historical anecdotes. Worcestre was also interested – more unusually – in topography, and particularly in buildings. The *Itineraries* are full of remarks on churches, castles, bridges, and houses, all recorded with an eye for the physical fabric. He provided measurements for most of the buildings he visited, either by pacing them himself or by interviewing resident craftsmen. In this respect, his notes provide an indirect witness to an otherwise elusive segment of medieval society.[9]

In late August 1480, Worcestre saw St Stephen's Church in Bristol, where he was particularly taken by the 'ingenious workmanship' of the south portal. He even provided a small sketch of its elaborate footprint, after having evidently fallen into discussion with the mason, whom he identified as Benedict Crosse (figure 9, catalogue 4). The sketch is something of an oddity: rather than a 'view' of the portal – as we might expect in a tourist's journal – it is a technical drawing of its cross-section. That circumstance and the lack of any similar drawings in the notebook suggest that it was Crosse himself who drew it and who also supplied the names for each of the portal's composite mouldings, which Worcestre dutifully listed: fillet, casement, bowtelle, and ogee, repeated in different combinations. The drawing was probably based on the template that Crosse had been using to cut the stones for the jamb. If that is the case, it would be a rare survival. Only two other template drawings from medieval England are known. Drawn to full scale, they are found in a late fourteenth-century model book held in the Pepys Library at Magdelene College, Cambridge (figure 10). Fortunately, St Stephen's Church still stands, and a comparison with its southern doorway shows Crosse's sketch to be reasonably accurate. He appears to have been able to replicate the portal jamb quickly and on a reduced size without distorting the proportions of its internal mouldings, which suggests that making and using such small-scale drawings was not unusual for him.[10]

The role of plans on paper or parchment in medieval England is obscure. They were no doubt employed throughout the Gothic period; L. F. Salzman has identified many instances in which 'plattes' and 'pictures' are mentioned in contemporary contracts. It is difficult, however, to know what these contract drawings looked like or how they might have been used, because virtually none survive. There are no English equivalents to the famous sketchbook of Villard de Honnecourt and scant few to the magnificent and detailed drawings from the German cathedral lodges of Strasbourg, Vienna, Cologne, Ulm, and Prague. It is supposed that such drawings once existed but, in the absence of a stable and conservative lodge tradition, they were lost through use, repurposing, or neglect. The lack of evidence has served to constrain interest in the topic, particularly in comparison to the wealth of research on continental drawings.[11]

Medieval Drawing and the Gothic Tradition

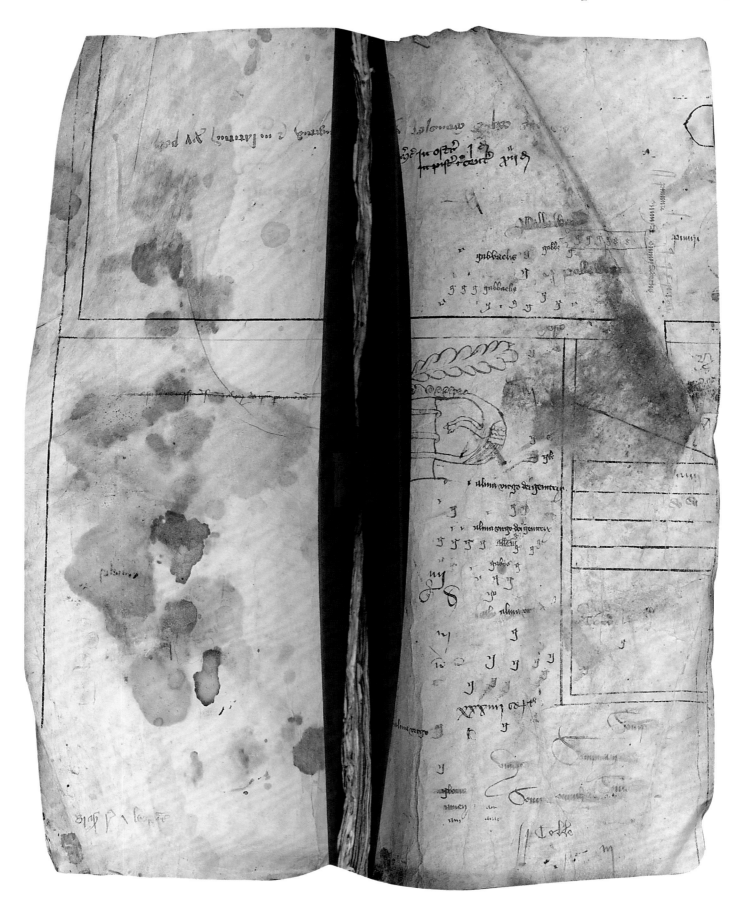

Compass and Rule

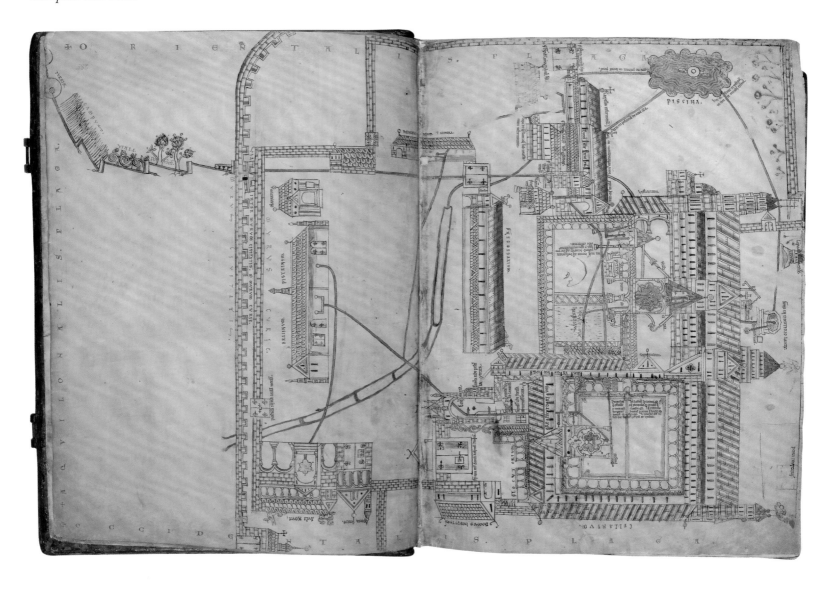

12. Canterbury Cathedral, 'Large Waterworks' from the Eadwine Psalter, *c.*1160. Trinity College, Cambridge, MS R.17.1, f. 284v–285r

Despite this handicap, there is general consensus about the purpose of paper plans, which appear to have differed from our own version of such documents, namely the 'blueprint'. Medieval plans may have been used to communicate the idea of both whole buildings and parts of buildings to patrons and workmen and to provide a binding record in a contract, but their utility as guides to construction was restricted in that they were not always measured drawings and did not serve to determine dimensions for use on the site. For the most part, this process had to occur on cleared ground, as the building was being laid out. In a sense, the detailed 'drawing' of plans took place on site using stakes and string, with many planning issues resolved not by erasing lines on paper but by moving stakes. In large masonry buildings such as churches, the composition was often worked up by the accretion of squares, with their rotated diagonals also used to determine certain proportions. These provided the basic modular units of the bays, as well as larger dimensions for the exterior envelope. The parties involved may well have consulted a drawn plan, not to be followed slavishly but only as an approximation of the geometrical procedures applied directly on the ground.[12]

The few surviving examples of plans from the period support this interpretation, for they are purely schematic. They exhibit no consistent scale but rather include room or wall dimensions

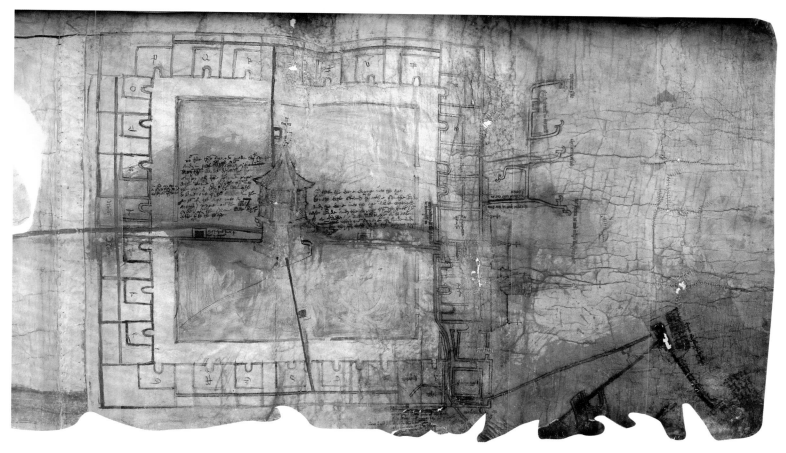

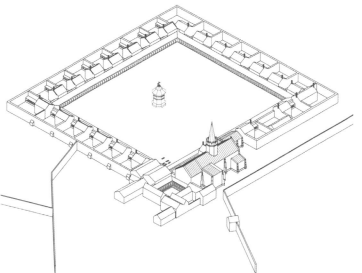

13. Water supply map from the London Charterhouse, mid-fifteenth century. London Charterhouse Muniments

14. Reconstruction of the Great Court of the London Charterhouse, from Barber and Thomas, *The London Charterhouse* (2002)

written out longhand and could not, therefore, have been intended or used as sufficient guides for construction. The earliest extant work of this kind is a partial plan of the court and kitchen of Winchester College, part of the west range of the inner court, completed in 1394 (figure 11, catalogue 5). The drawing illustrates a number of salient points about the use of medieval plans. It is, in the first place, a rare survival, having been preserved only because it was reused as the inner binding of a hall-book, used to record the names of college diners for the year 1415–16. It would be difficult to overstate the significance of this admittedly inconspicuous fragment: it is the only working plan of an English building that survives from before 1500. Second, the drawing's presentation is also unusual. As John Harvey has remarked, for a technical drawing produced in view of a building project, it is disconcertingly out-of-scale. Although an inscription clearly identifies the site as that of the new kitchen, 'locus noue coquine', the proportions of the staircase and the courtyard wall bear little relationship to that of the existing buildings. Nor are the doorways to the kitchen and chambers marked. It is possible that the drawing was merely a rough preliminary to be succeeded by a more accurate plan, although this is belied by the size of the fragment and by the care with which it was ruled. It is more likely that the final 'drawing' was laid out on the site itself.[13]

The lack of proportional relationships in a true project drawing is mirrored elsewhere. The use of pictorial conventions for depicting buildings was widespread, even in drawings that would otherwise be considered purely technical. The famous 'Large Waterworks' drawing of the Canterbury Cathedral precinct is among the earliest examples of this technique (figure 12). Dated to c.1160, it depicts the monastic buildings within the complex in flattened profile, their elevations hiding their ground plans. The lengths of the buildings relative to each other are roughly accurate, which suggests that they may have been paced, but the distances between them are often distorted, particularly toward the northern precinct wall, where the structures appear in a loose, notional arrangement. From a modern point of view, the lack of consistency is startling. Even a rough plan would have provided a way of approximating the length of the water pipes, but the drawing was not made for that purpose.[14] Another water supply map reflects a similarly cavalier attitude toward the representation of monastic buildings. The mid-fifteenth-century plan of the London Charterhouse, the terminus of a long sequence of parchment skins devoted to the course of the conduit, combines what now appear to us as a jumble of representational conventions (figure 13). The four ranges of the cloister are shown in combined plan and elevation, whereas lively perspectival effects are used for the chapel roof and octagonal conduit head, the latter clearly outsized. A comparison with a modern archaeological reconstruction of the site makes abundantly clear the artist's lack of ability or interest in reproducing scale relationships (figure 14).[15]

It is apparent that neither of these drawings was intended to communicate anything but rudimentary information about the size and location of buildings. What then was their value? In contrast to modern plans, they were produced not as a stand-in for, but as a supplement to, first-hand knowledge of the site. We find the same understanding of the plan in other documents, even in those intended to provide a graphic record of property boundaries. The four London plots drawn in the 1470s from the 'Small Register' of deeds compiled by the London Bridge House are very rare examples of drawn medieval surveys (figure 15, catalogue 6). The boundaries of each plot are inscribed with dimensions, some recorded to the nearest inch. The properties were directly measured, probably with a perch rod or a knotted rope. Also recorded are the names of the surveyors, their official titles, and the date of the survey. To orient the viewer, the authors note relevant landmarks, 'by the hye wey', and adjacent properties: 'the lond of the viker of depford'. The specialized vocabulary of the inscriptions – words such as 'patron' and 'platte' – suggest that the land was surveyed and drawn by building practitioners, as does the use of ruled lines and the occasional marks left by the compass. The inscription on the fourth plan testifies that it was 'moten' by Robert Whetely, Master Carpenter of London Bridge. Given the legal status of the documents, the technical competence of the surveyors, and the specificity of the measurements, it is again surprising to find that the plots are drawn diagrammatically. The explanation no doubt lies in the nature of the documents, which are not to be understood as

15. Plan and elevation of London plot and 'tenements', c.1475. London Metropolitan Archives. Catalogue 6

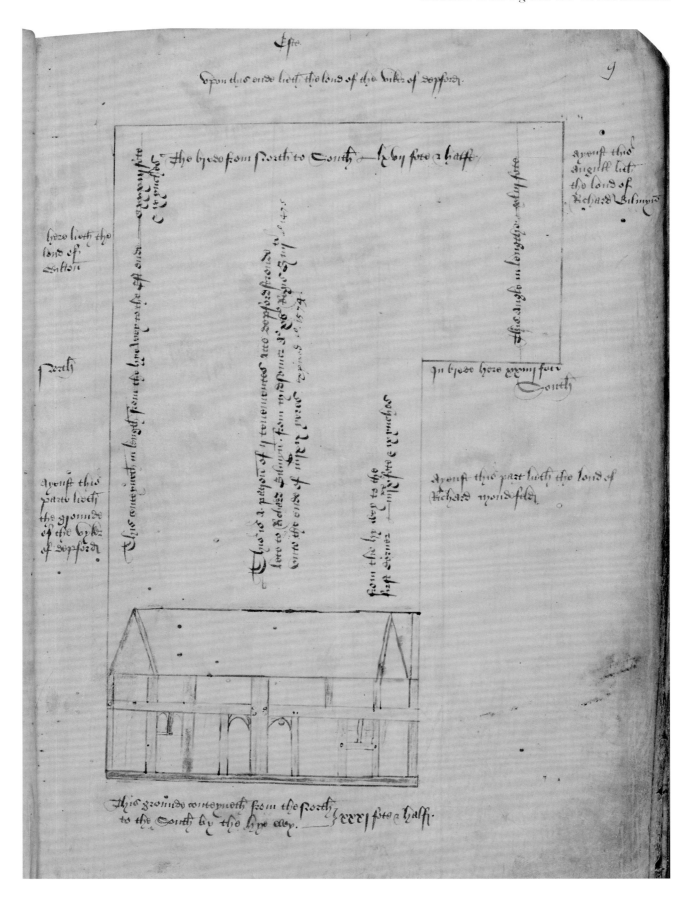

Compass and Rule

16. William Vertue (attr.), Proposal for a Tomb for Henry VI, c.1504–9. British Library, Department of Manuscripts.
Catalogue 7

17. William Vertue (attr.), Bay elevation of the chantry chapel of Bishop Richard Fox, Winchester Cathedral, c.1513–18. Royal Institute of British Architects, London.
Catalogue 8

graphic sources in themselves but as recorded witnesses to the official act of 'seeing', 'meting', and 'bounding' performed by the named officials of the Bridge House. Indeed, the inscriptions assume a familiarity with the land and neighbouring properties that would presumably make the use of scale superfluous. In this respect, the plans conform to most medieval surveys, which tended to be written, not drawn.[16]

The most common sort of medieval 'paper' drawings – and the closest medieval analogue to our own notion of architectural drawing – took the form of Gothic elevations. Although few English specimens remain, those that do survive are close in size, character, and quality to the great lodge drawings of the continent. Like their German counterparts, these external elevations are characterized by an emphasis on linear surface treatment, a casual use of perspective, and a profusion of precisely drawn decoration. The exquisite detail suggests that they were an early type of 'presentation' drawing, intended specifically for building patrons, allowing them to envision the completed project before construction. But they are also measured to a consistent scale and could conceivably have been used as guides in the building process. Two of the three included here have been attributed to the master mason William Vertue. The first is a proposal for a sepulchral monument to Henry VI, presumably made for Henry VII, who had planned to erect a shrine over his uncle's relics at Westminster Abbey (figure 16, catalogue 7). Phillip Lindley, who has dated the drawing to 1504–09, has discussed its iconography and style, which is characteristic of that used by Henry VII's royal masons.[17] The second elevation depicts a bay of Bishop Fox's chantry in Winchester Cathedral, erected sometime between 1513 and 1518 (figure 17, catalogue 8).[18]

Both monuments were intended to incorporate sculpture, and the designs include numerous niches and pedestals for this purpose. As Lindley has pointed out, the fact that no figures are included underlines the separation of the mason's role from that of the sculptor in late medieval England and the priority of the former in works calling for architectural design. In fact, both elevations are characteristic of the mason's art – of a style of drawing that is particular to the craft. Both are clearly 'constructed' with compass and ruler: the principal lines were ruled with a point and the circles of the major quatrefoils scratched in beforehand. An ink overlay was then set out on this basis, with minor lines and smaller details drawn freehand. The drawings also reflect the verticality and linearity of high Gothic architecture, with its emphasis on elongated structural supports. The treatment is correspondingly flat and, despite the use of light wash to highlight sculptural relief, there is little emphasis on volume or mass. The designer of the Henry VI monument has made some attempt to alleviate this particular quality of the drawing, by using an improvised perspective for the short side of the tomb. His discomfort with the technique, however, is apparent, particularly in the cusping and the joins between the superstructure and the octagonal buttress-posts. There is, finally, the same intensity of detail across all parts of the drawings. The craftsman was evidently not trained to think pictorially in terms of the design as a whole, but rather to divide it into a series of individual parts. These works may indeed have been presentational, but they

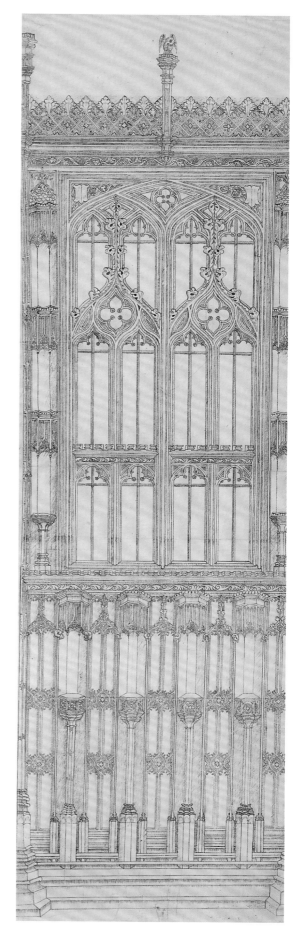

18. Elevation and combined plan of a canopied niche or pedestal, unknown date. Royal Institute of British Architects, London. Catalogue 9

nonetheless derive from an ingrained habit of using and producing technical, working drawings.

The third example of this type exhibits the same qualities, but with an additional element that is also characteristic of the mason's practice. The drawing depicts a pedestal for a sculpture surmounted by an elaborate free-standing canopy, intended either for a niche of a reredos or chantry or as a free-standing monument (figure 18, catalogue 9). Uniquely among the few surviving architectural drawings from medieval England, it unites elevation and plan in the same view, with the elements of each aligned on the page and drawn to the same scale. The plan shows three stages of the canopy superimposed, one within the other, with the crown at the centre. The technique is a staple of Gothic drawing. Not only does it make legible elements of the elevation that project towards or recede from the picture plane, it also reflects a method of generating the design of the elevation itself, which was typically developed by extrapolating measured elements of the ground plan. The 'combined view' would later be adopted as a visual convention by theorists of geometrical perspective, such as Albrecht Dürer and Piero della Francesca. Here, it is important to note its origin in the mason's everyday practice.[19]

Chapter Two
The Paper Revolution: The Origin of Large-scale Technical Drawing under Henry VIII

The first important transformation of medieval design practice occurred in a military context, a product of Tudor fortification engineers and their Italian counterparts brought into England by Henry VIII. Although they did not contribute to the spread of classical forms in Britain, military engineers did provide an important professional model for later generations of architects, in the modern sense of that term. Medieval building was carried out by teams of craftsmen working in loose concert but each within their own sphere of expertise. Even when acting as the designer, the master mason or surveyor in charge of a project was not expected to forecast and control the entirety of the building process in a way that is common for architects today. Nor did drawings have the same pervasive authority as a means to direct construction. Many design decisions were still made on the spot, by individual craftsmen during the course of work. This approach held most often for small-scale elements such as furnishings, decoration, and details, but it could sometimes extend to the design as a whole. It was not unusual even for major churches to be begun without a coherent plan for their completion. Indeed, many such buildings bear evidence of cumulative design changes, sometimes occurring over the course of generations.

In the realm of military architecture, cannon warfare rendered this mode of practice increasingly obsolete. Unlike most medieval building, bastioned fortifications required the subordination of individual craftsmen to a single authoritative design. In principle, each separate part of the fortress had to serve another. This was particularly true of the angle bastions, the size and position of which were governed by geometric necessity. The gun emplacements hidden within the neck of each provided cover for its neighbour, and together they created crossing fire against opposing batteries and the approach of miners. The elements of the system were mutually supporting, leaving no blind spots or dead ground anywhere within cannon range. This design idiom, known as the *trace italienne*, came to England comparatively late. The first angle bastions proposed for an English fort appeared in 1545–46 at Tynemouth, Portsmouth, and in the English possessions near Calais. A period of experimentation and adaptation followed, but by the time the powerful fort at Haddington was built in 1548 the basics of the new technology had been fully assimilated.[1]

The new fortresses allowed less scope for improvised design changes, not only because each element depended on another but because they had to be built quickly, in periods of urgency and likely attack. As a result, military engineers were among the first building practitioners to plan large-scale projects as coherent entities and with overarching authority. Their role in the building process was novel in another way. Tudor engineers were itinerant officials of the royal building works, who were frequently called away to respond to the defence of other strongholds or to report to the royal councils. The execution of the design was thus often left to others. It

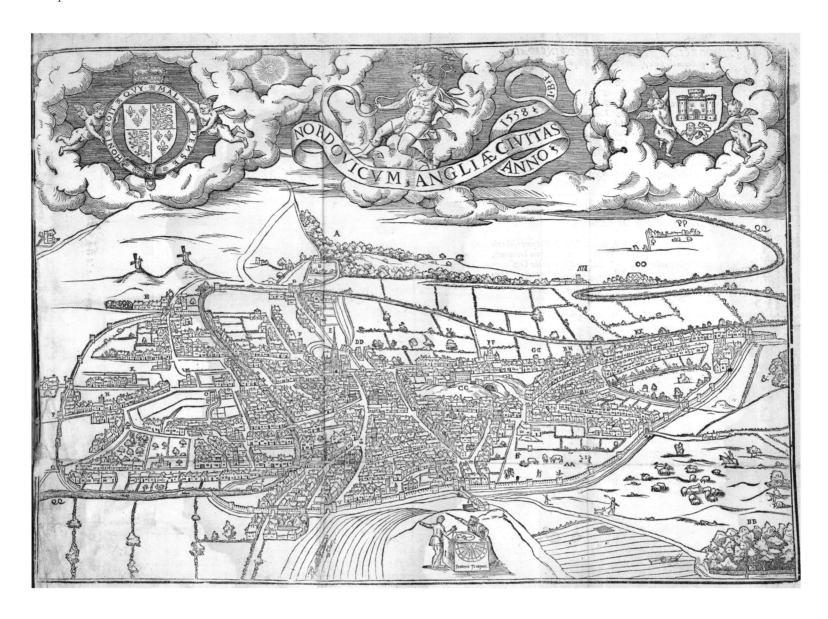

19. View of Norwich, from Cuningham, *The Cosmographical Glasse* (1559)

was in the realm of fortress-building that the first tentative steps toward the separation of design and construction – a defining attribute of modern architectural practice – took place. Contrary to modern expectation, these developments were not pushed forward by Italian engineers. Although such men were not uncommon in Henry VIII's military service, the pivotal figures were English: Richard Lee and John Rogers.[2]

This situation was reflected in two decisive advances in the planning process. The first was the more routine adoption of paper drawing itself, both as a medium of design and as a method of communication between patron and designer. These innovations took place in the Royal Office of Works during the 1540s, as the institution responded to the need for extensive modern defences in view of a very real threat of foreign invasion. In this context, technical drawings or 'plats' became the principal means by which the Crown's agents conveyed practical and strategic information from the site, and by which the royal councils – often under the

The Origin of Large-scale Technical Drawing under Henry VIII

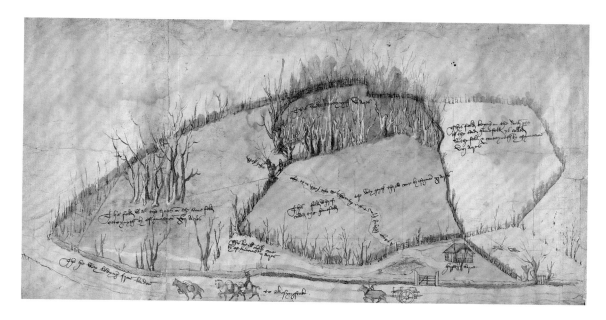

20. View of fields in Newnham, Hampshire, mid-sixteenth century. Winchester College Muniments, no. 3233

21. Proposal for a new harbour at Dover, Vincenzo Volpe, 1532. British Library, Department of Manuscripts, Cotton, Augustus I.i.19

personal oversight of the sovereign himself – made informed decisions and commissioned the necessary works. The treasure-trove of sixteenth-century maps and plans at Hatfield House and in the Cottonian collection of the British Library testifies to this effort. The items discussed here are drawn from the holdings of the latter collection.

A second revolution accompanied this wave of building: the introduction of new technical conventions in the drawings themselves. Our study pays particular attention to the earliest appearance of the scale building plan and the related technique of geometrical survey.

The traditional manner of representing towns was by means of 'chorography', typically an elevated view of a settlement in a wider landscape. The term had originated in Ptolemy's *Geographia* to distinguish world maps from the representations of specific places and regions. As William Cuningham defined it, chorography 'sheweth the partes of th'earth, divided in

Compass and Rule

themselves. And severally describeth, the portes, rivers, havens, fluddes, hilles, mountaynes, cities, villages, buildings, fortresses, walles, yea and every particular thing, in that parte conteined'. Whereas geography utilized the abstract geometry of the sphere, this science was understood to appeal primarily to the senses: 'as if a painter shuld set forth the eye, or eare of a man, and not the whole body, so that chorographie consisteth rather in describing the qualitie and figure, then the bignes, and quantitie of any thinge'.[3] Cuningham illustrated the concept with a double-spread engraving of Norwich, 'as the forme of it is, at this present 1558', with a legend of the city's landmarks, principal streets, city gates, and neighbouring villages (figure 19). It is, in fact, the earliest printed map of any English town. As in this example, a chorography might be built up from an initial measured survey – notice the author and his assistant standing over an outsized magnetic compass in the foreground – but, for the most part, this type of pictorial 'map' lacked any consistent scale or standardized technique. Such scenes were typically composed by uniting fragmentary views from several different vantage points to form an impressionistic whole.[4]

The scholarship on this subject has focused on engraved town views of the kind illustrated by Cuningham and compiled in Georg Braun and Franz Hogenberg's *Civitates orbis terrarum* (Cologne, 1572–1618). These collections were aimed primarily towards armchair travellers and humanist libraries, but they also appear to have penetrated to the level of the craft practitioner, who began to adapt the form to land surveys and large-scale project drawings from the early sixteenth century. One of the earliest known examples of the type is a bird's-eye view of fields in the parish of Newnham, Hampshire, made sometime between 1536 and 1551 for a lawsuit concerning the property (figure 20). It belongs partly to the tradition of the medieval picture map – the figures and the trees, for example, are far too large relative to the area depicted – but there is also a remarkable coherence to the image, the product of a real attempt to imagine the area from an oblique, elevated position.[5]

An analogous example comes from the early history of Dover harbour. Vincenzo Volpe's 1532 scheme, commissioned by the town in an attempt to secure government aid, was one of the first of a long line of proposals made for the site through the course of the sixteenth century (figure 21). Volpe's watercolour envisions two inner harbours dredged from the widened mouth of the Dour, the seaward one shielded by wooden jetties projecting into the bay. The drawing presents a curious mix of pictorial conventions. It is, in the first place, a 'technical' drawing for a large-scale engineering work, evidently intended to illustrate the wooden construction of the jetties, shoring, and sluice gates. Yet, the drawing is also conceived as an urban chorography. The visual tropes of the genre – church steeples, the town walls, and the castle in the background – are haphazardly emphasized, to the detriment of any consistent scale. If that were not enough, Volpe has also given the scene the air of a spectacle, adding ships sailing into the harbour and cannons sounding out from fortified towers. The image was, after all, intended for the king's eyes.[6]

One might expect such a pictorial showpiece from Henry's court painter. It is more surprising to find one from the hand of a royal mason and fortification engineer. A beautiful view of Calais harbour *c*.1541 has been attributed to Richard Lee, Surveyor of the Works in the Calais Pale (figures 22, 23, catalogue 10). The drawing is one of two of the site by the same hand. Both are related to fortification projects initiated in 1539, following Henry's excommunication and the ensuing alarm over a potential Catholic alliance against England. The drawing shows a proposal for two new gun towers on the cliffs east of the city, defending the approach from the sea. In view of its practical purpose, it includes most of the necessary information for the king and his council to make a decision, including the towers' relative proximity to Calais, the length

22, 23. Richard Lee (attr.), view of the Town and Harbour of Calais, *c*.1541. British Library, Department of Manuscripts. Catalogue 10

Compass and Rule

24. John Rogers or Richard Lee, Survey plan of Guines, c.1541. British Library, Department of Manuscripts, Cotton, Augustus I, Supp. 14

25. Anonymous Portuguese engineer, plan of town and castle of Guines, 1541. This plan is oriented 90 degrees clockwise from Rogers' above. British Library, Department of Manuscripts, Cotton, Augustus I.ii.23

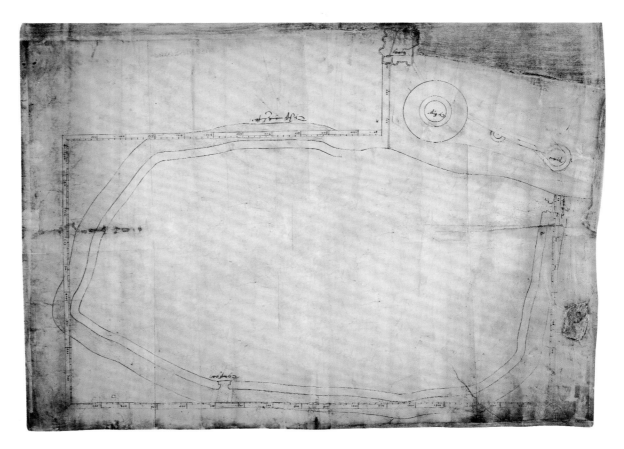

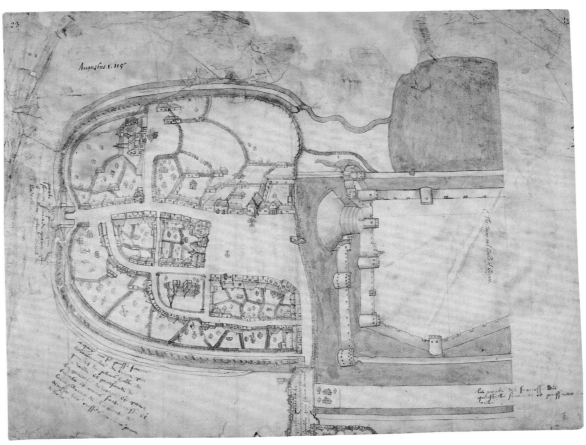

The Origin of Large-scale Technical Drawing under Henry VIII

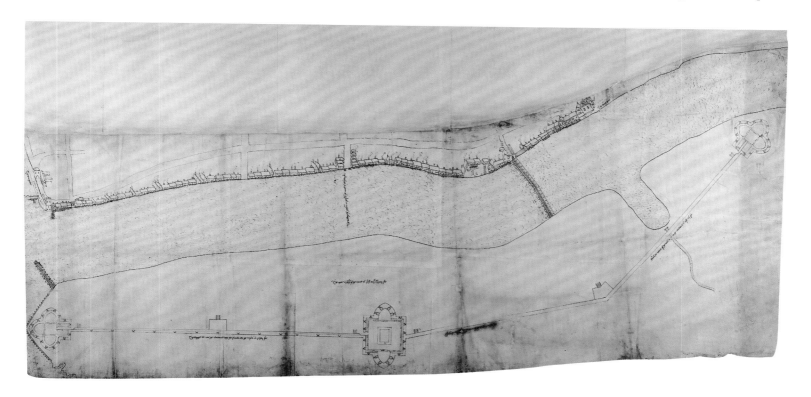

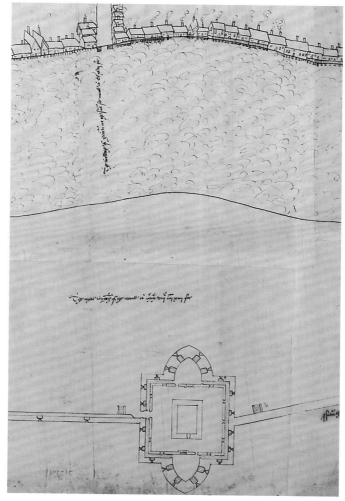

26, 27. John Rogers (attr.), plan for fortifications at Hull, 1541. British Library, Department of Manuscripts. Catalogue 11

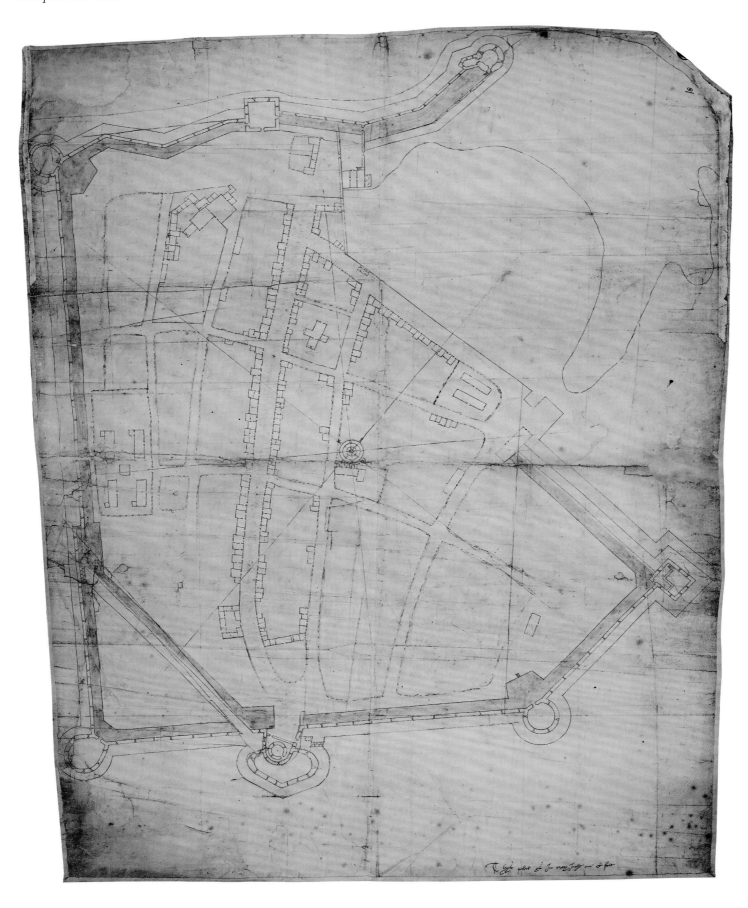

of the wooden jetties at the mouth of the harbour, and the outline of navigable sea lanes, marked out with pink wash. Yet what stands out for modern eyes is that the proposal has been conceived pictorially. It is, in essence, a highly competent landscape, enlivened with picturesque flourishes, such as the red roofs of the houses in Calais and the vertical hatching of the cliffs on the shoreline. Even more than Volpe, the artist took particular delight in representing ships, which are drawn in exquisite detail.[7]

For all their visual charm, chorographic views necessarily lacked precision, particularly regarding distances. Without the help of an interpreter with first-hand knowledge of the site, this handicap could present serious obstacles for decision-makers. This was especially true for fortifications, which were largely determined by lines of fire and the range of cannons. These circumstances must partly explain the appearance of the first English-made scale plans and geometrical surveys.

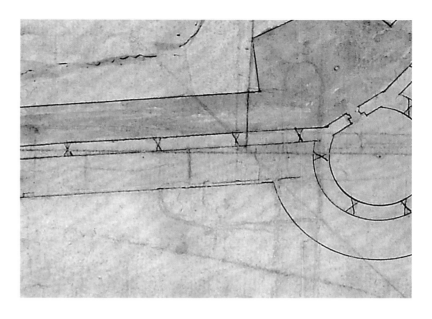

29. Detail of figure 28, pencilled addition to plan, showing proposed bastion at north round tower

Although scale maps of towns and large engineering projects had been used in Italy since the end of the fifteenth century, the technique remained virtually unknown in England. The earliest surviving examples occur in a group of four plans or 'plats' probably made in early 1541 by Richard Lee and the mason-engineer John Rogers at Guines, another of the English strongholds in the Pale of Calais. A preparatory survey of the existing walls and moat of the town can be considered the first of the series (figure 24). There are two unusual features of the plan that suggest that the concept of scale was new to the draughtsman or to his audience: a continuous graded ruler ringing the entire town and an inscription that records the scale in a roundabout way: 'the Inshe conteynyth L fotte'. Although it has none of the visual interest of a bird's-eye view, the plan represents a great conceptual and technical leap. Rather than mimicking the impression of an observing subject, it represents abstractly and indirectly the totality of physical measurements of the site, in a way that can be experienced by no single individual. Here, scale allows the drawing to become not simply an imagined 'view' of the proposed works but a mathematical instrument for both its planning and its realization. A contemporary drawing of the same site, attributed to a Portuguese engineer in the King's service, illustrates how novel the technique must have seemed (figure 25).[8]

This contrast is deliberately employed in another plan by Rogers, a magnificent presentation drawing for new defensive works at Hull in Yorkshire (figures 26, 27, catalogue 11). Completed in October 1541, it is linked to his work at Guines as part of the same nationwide scheme of fortification, in this case to defend against a possible incursion by the Scots, newly allied with the French. Over six feet long, the drawing shows three proposed bulwarks or 'blockhouses' east of the town, connected by a fortified wall almost 2,500 feet in length. The entire complex was to run along the river, opposite the town. This drawing and two others for the same project happen to be the earliest known scale plans for architectural works in England, but the document's interest goes beyond this fact of chronology. Even at this incipient stage, Rogers has adapted the technique to a rhetorical purpose, not only by drawing the plan to such a large size, but also by contrasting it with the small houses across the river. Roughly sketched in bird's-eye view, these are even embellished with charming touches, including gabled roofs, little doors, windows, and

28. Survey plan of Portsmouth, 1545. British Library. Catalogue 12

Compass and Rule

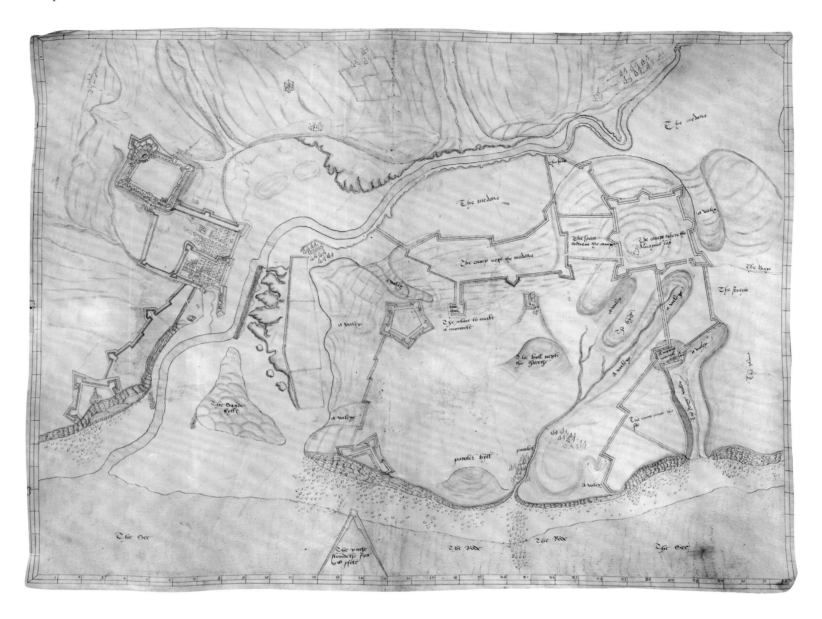

30. John Rogers (attr.), topographical survey of Boulogne with English counter-fortifications. British Library, Department of Manuscripts, Cotton, Augustus, I.ii.53

puffs of smoke from the chimneys. The plan of the bulwarks, on the other hand, is draughted in delicate, precisely ruled lines. The contrasting styles call attention to the project's massiveness, to its defensive strength, and – not least – to the meticulousness of its planning.[9]

The scale of these drawings approaches that of local maps and city plans, so it is perhaps not surprising to find Henry's programme leading to innovations in that area also. The famous Portsmouth Map of 1545 is the first 'ichnographic' plan of an entire English town, in which all of the topographical features – from the proposed fortifications to small dwellings – are represented in plan and to scale (figure 28, catalogue 12). Indeed, the absence of any pictorial imagery or out-of-scale elements is striking. In this respect, the map reflects the geometrical procedures used to make it. The basic technique was to make a measured transit around the town walls, recording changes in orientation with a magnetic compass. For greater accuracy, the map-maker would then have correlated the results with a second set of readings, made by plotting topographical features from a fixed point, using a radial sighting instrument. The wind rose in the centre of the drawing may represent this process. The cartographic conventions used here have become so ubiquitous that it is difficult to appreciate the plan's originality. In fact, maps of this kind were rare even in Italy. Other than Leonardo da Vinci's plan of Imola of 1502 – the earliest known instance – few examples survive from anywhere in Europe before the middle of the sixteenth century.[10]

The plan's authorship is unknown, although Richard Lee may have had a hand in it. He had been placed in charge of the town and its fortifications on 26 July 1545, following the events of 19 June, in which a French attack and landing on the Isle of Wight drew attention to the weakness of the local defences. The plan constitutes a response to this incident, combining a survey of the existing walls with a proposed improvement. The principal recommendation is shown in the walled area in the north-eastern part of the town, which is cut off by a transverse ditch and earthen rampart commanded by the great pointed 'bastillion' at the end of the high street. The proposal is noteworthy because it introduces concepts of modern Italian engineering into what was essentially a medieval defensive system of high walls and tall, salient towers. There are, in fact, further alterations, sketched out in pencil, that develop this strategy. Two Italianate angle bastions have been added to support the great 'bastillion', one at the point where the south-east wall joins the traverse and another immediately adjacent to the round tower north of the town (figure 29). To this latter element is attached a ruled line indicating another transverse rampart leading toward the harbour and further reducing the defensive circuit. Additional pencil marks show how the flankers located in the neck of the 'bastillion' might be adapted to rake the ditches on either side. Although the alterations were not carried out, their presence is nonetheless remarkable. As Martin Biddle has argued, the amended scheme is the earliest known plan for the defence of an English town by means of a fully flanked, bastioned system. In a more general sense, the pencilled additions illustrate the crucial mediating role that paper plans had come to fill among engineers and decision-makers. This was clearly a working document, which appears to have been the object of considerable discussion.

As an urban map, the Portsmouth plan also stands out for its scale – 'Thys plat ys In every Inche on C fote' – which is far smaller (as a ratio) than that of most sixteenth-century architectural drawings. In this respect, the plan fits within a broader shift in the history of English cartography toward the first large measured surveys. John Rogers was also at the forefront of this development, as in so many other advances in technical drawing. Among his plans for the defences at Hull, for example, is a map of the countryside around the town, showing settlements on both sides of the Humber, as well as roads, landmarks, and watercourses. Along with two drawings that he made of the area around Boulogne in 1546, they are the first English-made local maps to include scale bars. As a group, they were made for a single purpose, namely to situate large fortification projects in a topographical context. This circumstance is surely significant, for it suggests that the kind of strategic planning involved in fortification design lent itself not only to the use of scale plans but also – and almost immediately – to scale mapping. In this respect, the second Boulogne drawing is characteristic, for it shows an elaborate scheme of English counter-fortification in view of a potential attack on the French-held forts opposite the river Liane (figure 30).[11]

This shift must have initially imposed itself as a matter of military necessity, but the advantages of scale maps for other situations were quickly recognized. The first measured estate plans from the 1570s and 1580s, for example, are the direct descendants of the drawings produced by Henry VIII's engineers. To be sure, the chorographic tradition persisted for a long time.

Compass and Rule

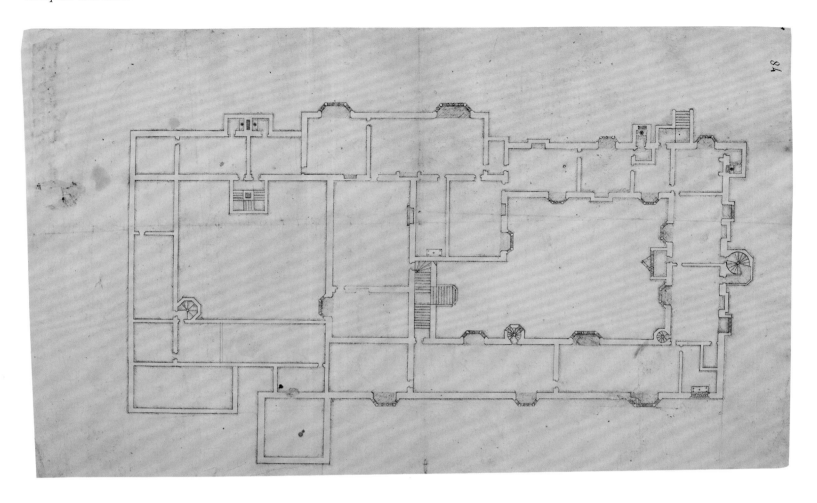

31. John Rogers, proposal for remodelling the first floor of Hull Manor, 1542 or 1543. British Library. Catalogue 13

Pictorial imagery continued to be used for maps as well as for large building works well into the seventeenth century. But the benefits of scale plans would prove increasingly compelling. In time, the technique would become a habit of thought so ingrained as to seem almost natural.[12]

Rogers himself appears to have appreciated the wider benefits of scale drawing almost immediately. During his time at Hull in 1542–43, he was assigned to remodel the interior of Hull Manor for use as a royal residence and administrative seat. Among the drawings that he produced for this commission are the earliest scale plans we have for a civic building in England (figure 31, catalogue 13). The engineer accompanied these plans with a large, bird's-eye view of the same house, probably intended as a presentation drawing to inform the king of the building's general situation (figure 32, catalogue 14). The contrast between the plans and the view is instructive, not because of the inconsistencies in perspective – which might be expected from a mason – but because the proportions of the layout have been purposely distorted. The wings drawn as diagonals – including the great hall separating the two courtyards – have been lengthened in relation to those before and behind. These elements were presumably altered to make the courtyard elevations more legible, but it is nevertheless striking that Rogers could depart so casually from the scale plan. He evidently did not expect the discrepancy to confuse his patrons or, indeed, to matter much to them.[13]

The Origin of Large-scale Technical Drawing under Henry VIII

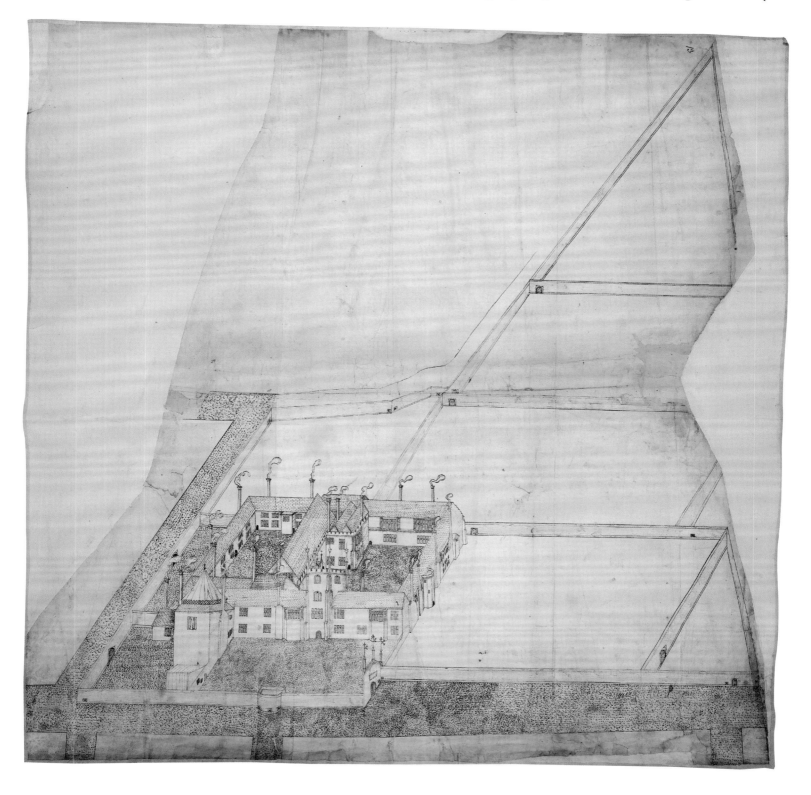

32. John Rogers (attr.), view of Hull Manor, 1542 or 1543. British Library. Catalogue 14

Scale building plans are attested from pre-classical times, but they do not appear to have been a tool of medieval building practice in England. One might simply point to the fact that none are extant from before 1540, but the nature and character of the earliest survivors is also telling. These first surviving scale plans emerged from situations where the need to portray conditions on the ground accurately, to maintain consistent proportional relationships in the layout of defences, and to provide a controlling authoritative design became – or were perceived to become – a matter of life and death. The fact that fortresses were generally too large to be designed on site may also have contributed to the adoption and spread of the technique. The simultaneous appearance of local maps drawn to scale provides further evidence of this phenomenon. P. D. A. Harvey has counted about forty existing measured surveys datable from 1540 to 1550, nearly all of them of town walls or castles, and probably all produced by military engineers.[14]

There remains one problem with this story, which historians have so far not been able to solve. How exactly did Lee and Rogers learn the use of scale drawing? It was long believed that they picked it up from itinerant Italian engineers. As Marcus Merriman has shown, there were several such men in the king's service. Yet, a possible point of contact is elusive. No Italian engineers, as far as we know, served at Guines, Calais, or Hull. More surprisingly, few of the existing plans from other sites that we can confidently attribute to an Italian author are drawn to scale. Of course, it is possible that English engineers discovered the technique themselves, as they grappled with the novel design problems involved in defending against cannon fire. Rogers had, after all, trained as a mason, and Lee's family also appears to have been connected to the craft. With such a background, it would have been a straightforward step to transfer the basic idea of scale from elevations and working drawings – which were common enough – to plans. Accurate surveying of towns and countryside was a trickier problem, but these tasks, too, would probably have drawn on a mason's normal training in land measuring. The duties of the craft often overlapped with those of the estate surveyor.

Whatever route Lee and Rogers took, the underlying cause of this shift remains the same. These men came from traditional craft backgrounds, but the crucible of war fundamentally altered their professional culture. New design challenges played a part, as did the demands of building at an unprecedented scale and with exceptional urgency. Moreover, the royal engineers were often thrown together not only with foreign engineers but also with practitioners in other crafts – from pilots to cartographers to gunners. All of these factors gave Henry's engineers a level of practical mathematical expertise that would have been very unusual among most civil builders.

Chapter Three
The Mathematical Practitioner and the Elizabethan Architect

In June 1542, the Captain of the English-held Guines Castle in the Pale of Calais was writing to the treasurer of the king's household in England. Before passing on more general news, he reported the arrival of four Kentish gentlemen and their request for the king's licence to travel elsewhere in Europe. For three days, the visitors had put on a sparkling display of their skills and made a remarkable impact on the garrison and their host. If only every county of England had such men,

> for thei haue amongiste them many good qualities aswell to do service by see [sea] as by lande and in thois thinges that thei haue reasenyd of here thei haue shewed themselfes marvelous wittely and thei cowde not be confowndyd by any that reasoned with them, aswell in gemetrie as thinges concernyng navigacon and the dissernyng of altitude as longetude, and as for the arte belonging to gonners I have sein none suche, in so muche that all thois that rekenethe theam selffes connyng on this sides of the see gevithe place vnto them, as well in argumentes of ther syences as in ther doing, whiche I haue bothe harde and seyn, not a litle reioysing thereat being gentilmen.[1]

Through the surviving plats of figures such as John Rogers, we have seen the design novelties being adopted by English technicians in France. This letter opens a window onto the wider culture that accompanied the English military presence. These four gentlemen clearly represented a novel and admirable phenomenon. Keen to serve the king, they were not just intellectuals but active intellectuals, with both a virtuoso technical mastery and practical abilities. Mathematics was their weapon of choice in the verbal jousting of informal disputations, with geometry as the foundation for their arguments and demonstrations in navigation, measurement, and gunnery. Even more surprisingly, they were gentlemen, in a world where the detailed grasp of such matters would more naturally be delegated downwards.

We do not know where these four men later travelled, or indeed if they left Guines at all. But we do know that one of them transferred the shared mathematical ethos of a small group of friends to the more public stage of print. Named simply as 'Digges' in the letter, the author can be readily identified as Leonard Digges: his experiments in artillery were still being recounted by his son decades later.[2] Digges was one of the first vernacular authors on practical mathematics in England, helping to shape the movement's early character and direction. He began publishing in the 1550s, a few years after Robert Recorde had composed his original series of English mathematical textbooks, ranging across arithmetic, geometry, astronomy, and algebra. Digges's first work was a highly successful almanac and he promised further works on subjects from cosmography to typography.[3] However, only one other significant text was issued

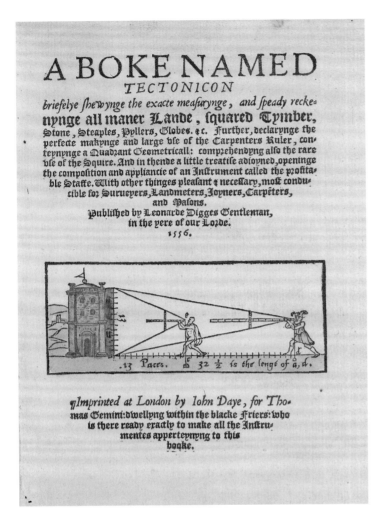

33. Title page from Digges, *Tectonicon* (1556). Bodleian Library, Oxford. Catalogue 15

34. Annotated title page from Digges, *Tectonicon* (1592). Bodleian Library, Oxford. Catalogue 16

in his lifetime: *Tectonicon*, a book on measurement, first printed in 1556.

The title page spells out not only Digges's subject but also his target audience (figure 33, catalogue 15). The work was aimed at the building and surveying trades – 'most conducible for Surueyers, Landmeters, Ioyners, Carpenters and Masons' – and it teaches the rapid and accurate measurement of plane and solid materials. Digges presents various instruments for measurement, from well-established craft tools such as the carpenter's rule and the square to a form of cross staff, which he christened the 'profitable staff'. Much of the book offers arithmetical rules and numerical tables to assist with the necessary calculations. Although focused on quantity and land surveying rather than the design and construction of buildings, Digges's work has a special place in the literature of architecture. In Eileen Harris's exhaustive study of British architectural books and writers, *Tectonicon* not only has the distinction of being the first volume in her chronology but was also one of the most successful and long-lived. There were eighteen issues in the period 1556 to 1656; the last was in 1692. Its sheer quantitative dominance is hard to overstate: it had more editions in its first century than all the more strictly architectural books in England put together.[4] Digges not only inaugurated but defined the tradition of practical mathematics as it related to building, and he provided a model for subsequent generations of mathematical practitioners.

Digges expected his books, including *Tectonicon*, to have a direct impact on practitioners. The work opens with a brief letter to the reader, whom he advises to look beyond the text itself. After taking up his books, readers should 'first confusely reade them thorow, then with more iudgement, and at the thirde reading wittely to practise ... Note, oft diligent reading, ioyned with ingenious practise, causeth profitable laboure'. Digges's injunction to diligence and ingenuity may seem unexceptionable, but it carried an edge. For Digges, the traditional practices of the crafts were insufficient and needed to be corrected by mathematical reason. He criticized as inaccurate the common rules for calculating quantities of stone and timber, and he substituted improved alternatives in their place. As both a substantial gentleman and a mathematician, Digges was not providing merely technical support but a prescription for a new relationship, in which the mathematician instructed the artisan from above.

Some self-consciously advanced artisans willingly accepted this vision, and vigorously sought to purge their art of 'vulgar errors'. In 1602 the joiner Richard More published *The Carpenter's Rule*, where he not only repeated Digges's criticisms but recommended other vernacular mathematical authors to his colleagues. He even advised them to attend the geometry lectures given at the recently established Gresham College. By echoing *Tectonicon*, More revealed that Digges's position did not command universal assent, and he recorded the protests of those who did not want to abandon their simple but erroneous rules in favour of mathematical exactitude. Who had the time and patience for this overly scrupulous approach when the

A BOOKE NAMED
Tectonicon,

Briefly shewing the exact measuring, and spedie reckoning all maner of Land, Squares, Timber, Stone, Steeples, Pillers, Globes, &c. Further, declaring the perfect making and large vse of the Carpenters Ruler, conteining a quadrant Geometricall: comprehending also the rare vse of the Squire. And in the end a little Treatise adioyning, opening the composition and appliancie of an Instrument called the profitable Staffe. With other things pleasant and necessarie, most conducible for Surueyers, Landmeaters, Ioyners, Carpenters, and Masons.

Published By Leonard Digges Gentleman, in the yeare of our Lord, 1556.

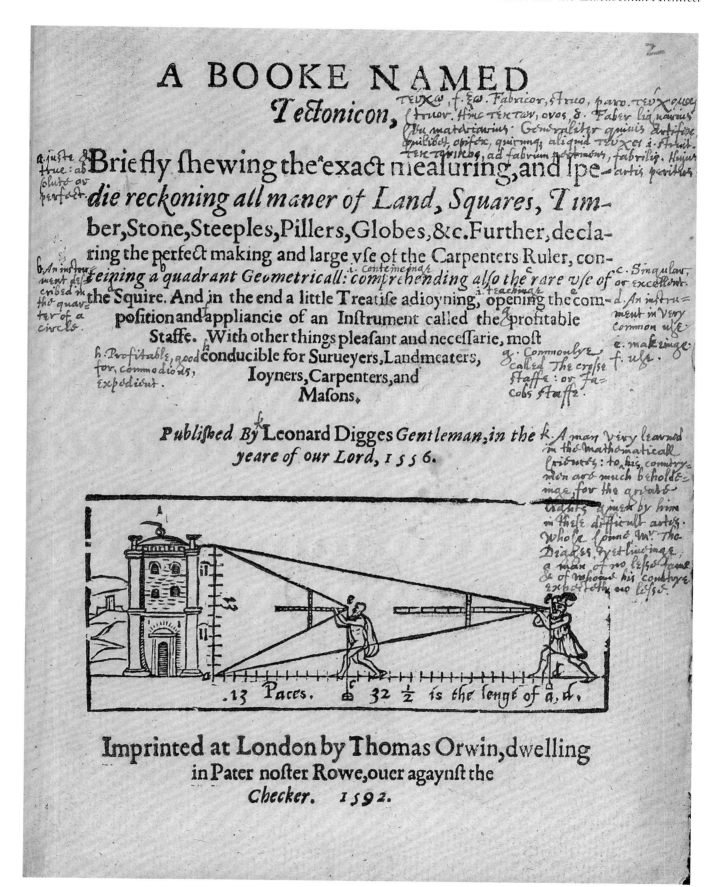

.13 Paces.   32 ½ is the lenght of a, d.

Imprinted at London by Thomas Orwin, dwelling in Pater noster Rowe, ouer agaynst the Checker. 1592.

35. Carpenter's rule, from Digges, *Tectonicon* (1556)

36. Line drawing of a carpenter's rule from the wreck of the Mary Rose, 1545

37. Reverse of Digges's carpenter's rule, from *Tectonicon* (1556)

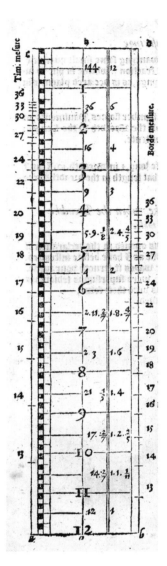

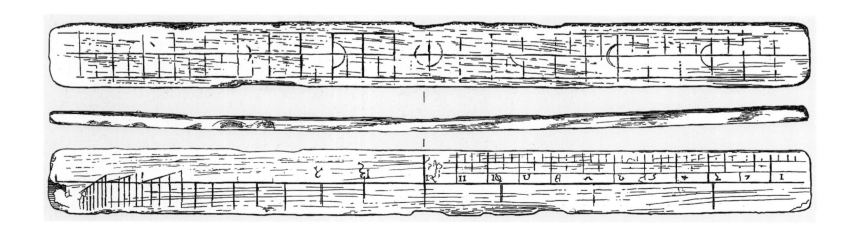

accepted rules were good enough? Craftsmen and merchants contended that allowance could always be made for any inaccuracies in practice.[5] For all its apparently straightforward procedures and sober self-evidence, Digges's intervention was not accepted without dispute. When the authority of the early modern trades was challenged, even quantity surveying could become a matter of controversy and resistance.

The response to *Tectonicon* from craft practitioners was therefore complex. But we can also see that at least some copies came into the hands of readers whose predilections and interests were far from artisanal. A copy of the 1592 edition in the Bodleian Library is an extraordinary testimony to the ways in which books could travel beyond the author's expectations and intentions (figure 34, catalogue 16). The title page and the early sections of the book are densely populated with marginal notes. These annotations form an unintentionally hilarious *explication du texte*, treating the work not as a route to practice but as an end in itself. The term 'Tectonicon' is glossed in Latin, but the reader noted its Greek root and used letters to identify other catch-words on the title page. 'Exact', for example, is marked with an 'a', which is picked up in the margin as 'a. juste & true: absolute or perfect'. A few lines on we find quadrant unpacked as 'b. An instrument described in the quarter of a circle', immediately followed by 'conteininge' as an elucidation of 'comprehending'. The author of this somewhat pedantic exposition is unknown but was certainly writing just after the publication of the book. When Leonard Digges's name is reached on the title page we are up to the letter 'k' and the marginal comment: 'k. A man very learned in the mathematicall sciences: to [whom] his countrymen are much beholdeinge for the greate lightes giuen by him in these difficult artes whose sonne Mr Tho. Digges yet liueinge a man of no lesse fame & of whome his country expecteth no lesse'. Thomas died in 1595 and our scholiast shows that, whatever Leonard had intended, in less than forty years of publication, the practical mathematics of *Tectonicon* had found a much broader audience than craftsmen alone.

Texts provide one way to assess the audience and the impact of the mathematical programme, but there is also another. *Tectonicon* is centrally concerned with the construction and uses of instruments. While Digges reprimanded craftsmen for their false methods of calculation, he also castigated the inadequacy of common measuring rulers. As More and other authors explain, inaccuracies were introduced and spread by the process of manufacture, since craftsmen typically made their rulers by copying another.[6] But just as mathematical practitioners provided new methods and tables for arithmetical calculation, so they could also supply improved instruments. Again, it is *Tectonicon* that announces this shift. The title page of the first edition concludes: 'Imprinted at London by Iohn Daye, for Thomas Gemini: dwellyng within the blacke Friers: who is there ready exactly to make all the Instrumentes apperteynyng to this booke'. Gemini, an immigrant from the Low Countries, was not only a publisher but an entirely new kind of figure in London life: a skilled engraver, responsible for the earliest significant copper-plate prints in England, and a commercial maker of mathematical instruments.[7] His products promised the accuracy that the text promoted. Artisans were now no longer advised to make their own instruments but invited to become customers of a specialist retailer.

Gemini and his successors had no more inevitable a path to success than the authors with whom they collaborated. Their authority and standards had to be established just as painstakingly, and in the face of the same craft independence and resistance. Over the century of *Tectonicon*'s textual dominance, instruments were continually developed and redesigned to place the message of mathematics quite literally in the hands of craftsmen. These instruments suggest the complexity of the process by which mathematical practice was accepted as a trusted resource by building practitioners. We can follow that process through the changing form of one

Compass and Rule

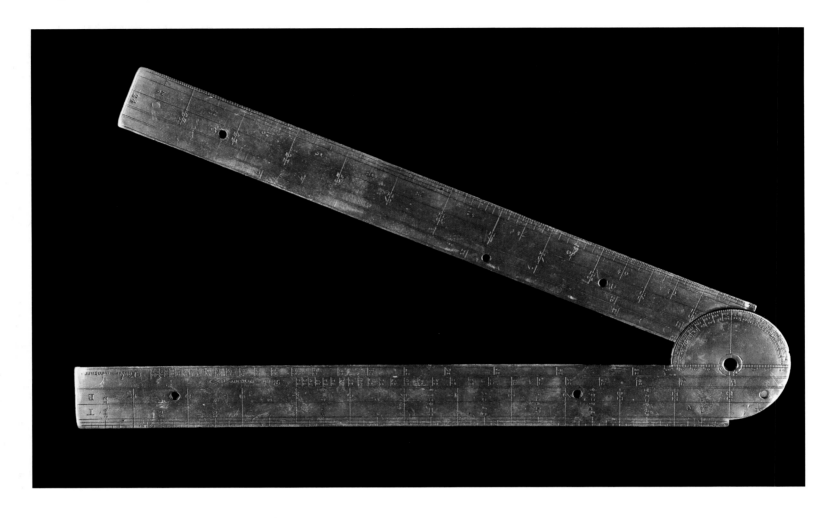

38. Humfrey Cole, surveyor's folding rule, 1575. Museum of the History of Science, Oxford. Catalogue 17

particular device, the carpenter's rule that Leonard Digges discussed in *Tectonicon* (figure 35). This daily tool of the trade, 'well known and commune amonge good Artificers', was both a measuring and a calculating device. Digges provided his own version, illustrating a straight rule one foot long (though 'truly it were more commodious, if it had two fote in length') with a scale of twelve inches on its face.[8] On one edge was placed a scale of 'timber measure', used to work out the volume of materials, and on the other a scale of 'board measure', for surface areas. This represented at most a correction and refinement of the traditional form of the device. The carpenter's rule found among the tools on the wreck of the *Mary Rose* (1545), for example, is a similar two-foot wooden rule, with inch scales and the scale of timber measure (figure 36).[9] Digges, however, supplemented the design further, by adding a miniature quadrant to its reverse side, for quick angle measuring (figure 37). This was a familiar device, used in surveying and for taking heights and dimensions, but here its small size, compressed into the width of the ruler, made it little more than a gadget.

After Digges had transformed the simple carpenter's rule, his design was then transformed in turn, probably in an attempt to appeal to a wider market. The version produced by the London maker Humfrey Cole replicated the standard features of the device, such as inch, board, and timber scales, but presented them in a new, hinged format that endowed the standard two-foot

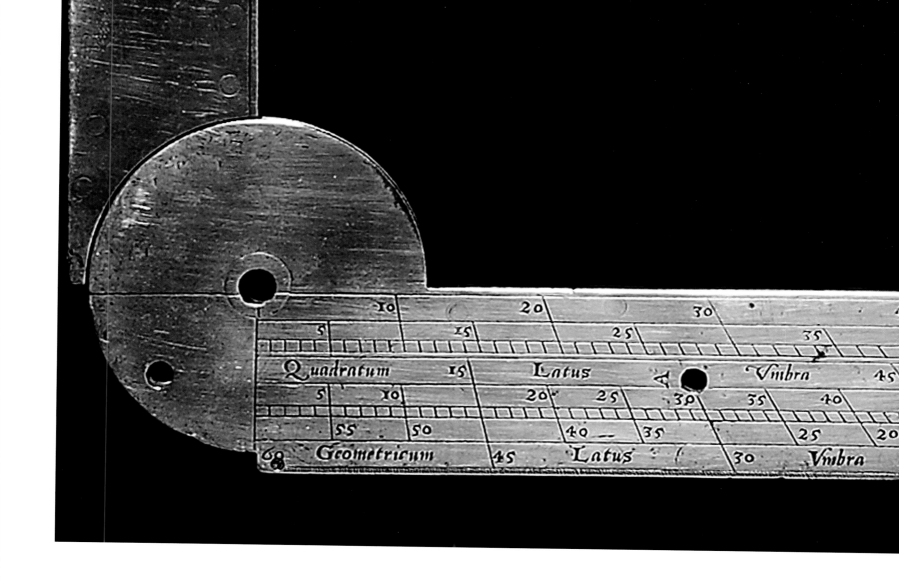

model with greater portability (figure 38, catalogue 17). In addition, locking the arms at right angles formed a quadrant of much larger and more effective dimensions than Digges's (figure 39). New features were added: the inches were variously subdivided to be used with different scales on maps and plans, and pairs of sockets were driven into each leg to receive sights, so that the whole could be employed as a surveying instrument mounted on a staff. While sober in appearance, this instrument must have been a striking symbol of the power and ingenuity of mathematics. What had started as a craft tool became, in Cole's hands, a near universal device of mathematical practice. One of the four surviving examples even incorporates a sundial.[10]

Cole's brass instruments would have been beyond the financial means of ordinary craftsmen. What we know of his customers demonstrates that his combination of innovation, accuracy, and restrained decoration appealed to a more elevated audience.[11] Thomas Gemini's productions were even more highly prized: of his seven surviving instruments, four have royal connections. His astrolabe of 1559, for example, carries the title and arms of Elizabeth I and was probably commissioned by Sir Robert Dudley (the future Earl of Leicester) for presentation to the Queen (figure 40, catalogue 18). Particularly in a royal context, the astronomy and time-telling of the astrolabe may seem a world away from the mundane measurements of *Tectonicon*. Yet, as well as being produced by the same makers, the functions of the instruments themselves overlapped. Elizabeth's astrolabe, like most others, carries a so-called 'shadow square' that allows the

39. Humfrey Cole, surveyor's folding rule, reverse, showing quadrant and shadow square scales for use with a plumb line

measurement of heights and distances. Exactly the same scale appears as part of Digges's quadrant and its enlarged version on Cole's rule.

In producing such prestigious instruments, Gemini was clearly working at the pinnacle of the art, but what of the humbler end of the scale? The only instruments to survive from the first generations of London makers are in brass, but we know that some specialists worked in the less expensive material of wood.[12] Their products may not survive, but later versions can give us a sense of what they looked like. Seventeenth-century examples of wooden carpenter's rules sometimes carry naïve decoration that contrasts with the restraint characteristic of the upper layer of London's mathematical instrument makers, suggesting both different makers and a distinct audience (figures 41, 42, catalogue 19). Evidence for a more variegated market also emerges from London guild records, which show that rules of various types were available not only from instrument makers but also from ironmongers and cane sellers' shops. The Clockmakers' Company (to which many instrument makers belonged) kept records for the ritual breaking of inaccurately made rulers and, judging by these documents, the instruments available in those shops were likely to be at the lower end of the market.[13] That is not to say that everything made in wood was necessarily inferior. Devices carrying novel scales designed to further ease the labour of calculation and measurement were made in increasing numbers, often following the folding format introduced by Humfrey Cole (figure 43, catalogue 20).

Whether acquired by artisans or gentlemen, elaborate versions of the carpenter's rule became established as part of the makers' routine repertoire and began to move out beyond London. A fine brass double-folding rule, now in National Museums Scotland, shows how provincial England followed the metropolitan example (figure 44, catalogue 21). From its materials and engraving, the piece is surely London-made, and it carries the most recent of reckoning scales, as well as the still relatively advanced provision of a logarithmic line for calculation. Significantly,

40. Thomas Gemini, astrolabe for Queen Elizabeth, 1559. Museum of the History of Science, Oxford. Catalogue 18

41, 42. Wooden carpenter's rule, 1648. Whipple Museum, Cambridge. Catalogue 19

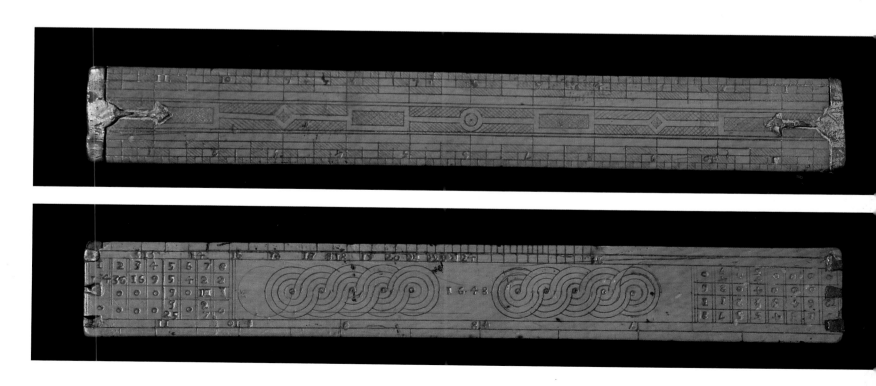

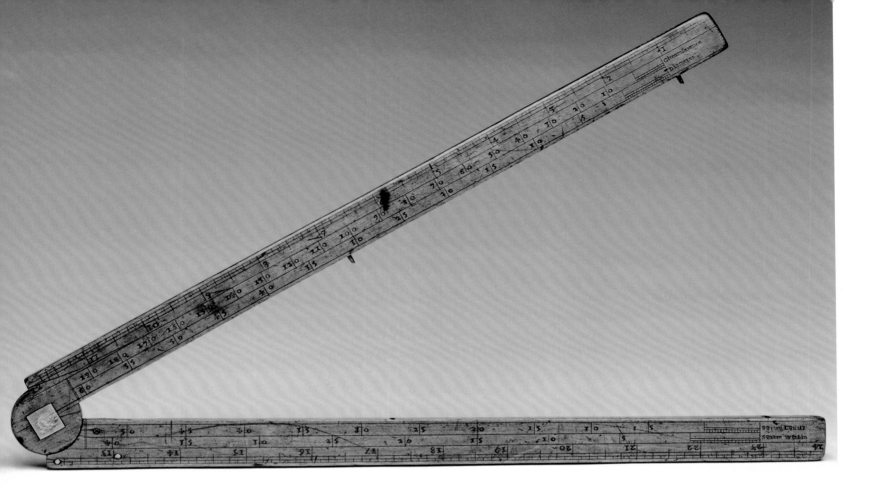

43. Wooden carpenter's rule, 1659. Science Museum, London. Catalogue 20

it is signed '1655 Robert Trollap of yorke free Mason'. Trollap, or Trollope, was the owner, not the maker. He is a documented builder and his rule is dated at a significant moment in his career. In 1655, the Corporation of Newcastle contracted with him to build their Exchange and Guildhall, 'according to the best Authors now in English'. Trollope not only undertook public, private, and church commissions but even carried out military work. His range harks back to that of the Henrician engineers and suggests how spheres that had long since separated and specialized in London and at court were still connected in the provincial realm.[14]

Trollope's folding rule suggests that, by the mid-seventeenth century, the mathematical programme was serving to define and bolster the identity of some successful building practitioners. Indeed, the process began much earlier; the earliest English writers on the orders made this connection explicit. John Shute, for example, could claim that architecture

> hath a natural societie and as it were by a sertaine kinred & affinitie is knit unto all the Mathematicalles which sciences and knowledge are frendes and a maintainer of divers rationall artes: so that without a meane acquaintance of understanding in them neyther painters, masons, Gold smythes, enbroderers, Carvers, Ioynars, Glassyers, Gravers, in all Manner of metals and divers others more can obtayne anye worthy praise at all.[15]

Robert Peake, too, linked the importance of sound geometrical knowledge to a basic competence in the art. In publishing a translation of Serlio's first book, he aimed 'to benefit the Publicke and convay unto my Countrymen (especially Architects and Artificers of all sorts)

The Mathematical Practitioner and the Elizabethan Architect

these Necessary, Certain, and most ready Helps of Geometrie: The ignorance and want whereof, in times past (in most parts of this Kingdom) hath left us many lame Workes, with shame of many Workemen …'.[16]

Programmatic claims of this sort are a commonplace of early modern architectural literature, but they do appear to have resonated in the actual careers of Elizabethan building practitioners. In a pioneering article published fifty years ago, John Summerson identified three figures in the royal Office of Works as architects in something like the modern sense, characterizing them as professional designers rather than the ordinary run of administrators and designer-craftsmen.[17] Mathematical practice was not Summerson's theme, but this perspective brings out significant points of connection between all three of his architects. Robert Adams, for example, was an expert in fortification, praised for his 'perfect skill in nombres and measures' and his ability to work 'by the exact rules of Geometricall observations'.[18] He fostered his abilities in collaboration with the mathematical instrument makers Humfrey Cole and Augustine Ryther.[19] The mason Robert Stickells was an artisan intellectual, able to cite Vitruvius, deploy the orders, and argue for the interrelatedness of proportions in building. His writings are obscure – 'semi-literate and at the same time esoteric', in Summerson's phrase – but he was clearly seeking to join theory and practice, with mathematics as a vital element of his aesthetic and professional identity.[20]

Summerson's third Elizabethan was John Symonds, a widely experienced craftsman, building consultant, and surveyor. He had been apprenticed to the mason and joiner Lewis Stocket, Surveyor of the Queen's Works, and was considered capable in a range of materials, being

44. Brass folding rule, 1655. National Museums Scotland, Edinburgh. Catalogue 21

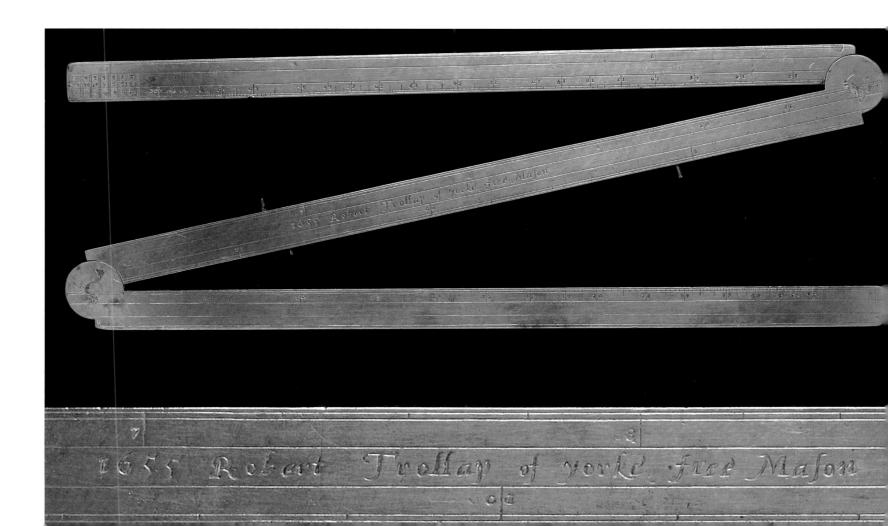

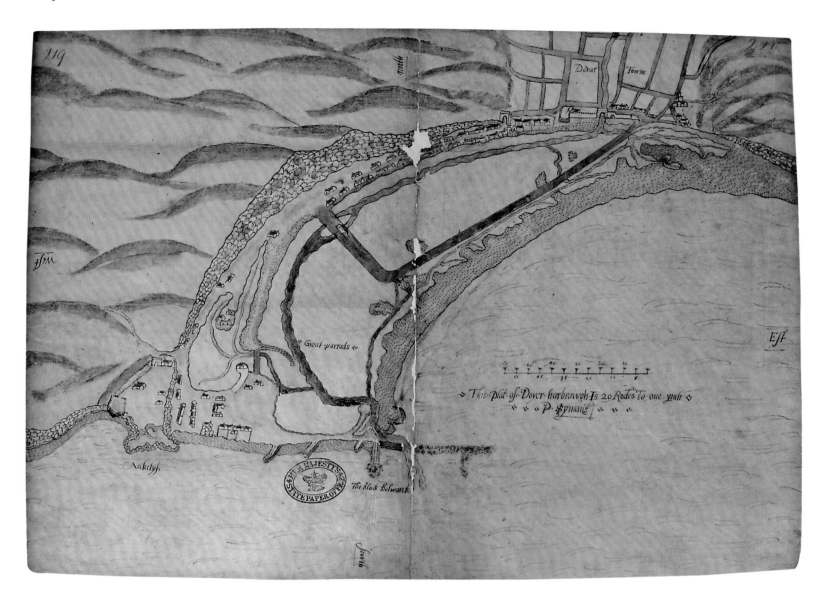

45. John Symonds, plan of Dover Harbour, 1583–84. National Archives, Kew. Catalogue 22

variously responsible for masonry work, stone carving, and carpentry. Symonds himself appears in the Works accounts as a clerk and purveyor, supervising and executing a wide range of projects from the late 1560s to the early 1580s. In 1585, he obtained the reversion of the office of Master Plasterer of the Works, holding the post from 1590 until his death in 1597. His official duties led to important private commissions, particularly for the Cecil family. He worked for Lord Treasurer Burghley at both Theobalds and Burghley House and advised his son Sir Robert Cecil.[21] Although primarily involved in architectural construction and decorative work, he was also consulted on technical matters in military and engineering projects. He was known to have designed an earthen gun-platform and, as Surveyor of Dover Castle, carried out repairs there in the early 1580s. He performed a range of tasks for the major harbour project at Dover. These began in 1583, when he advised on the construction of a masonry sluice, and he subsequently sounded depths within and around the harbour, drew up plans, and estimated quantities of materials. In the following year, he was called to confer on the design of jetties and

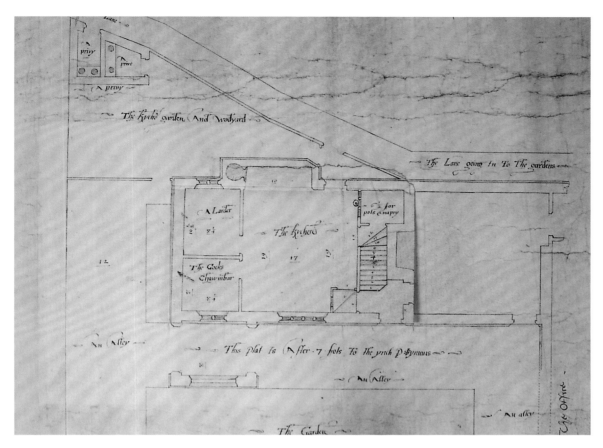
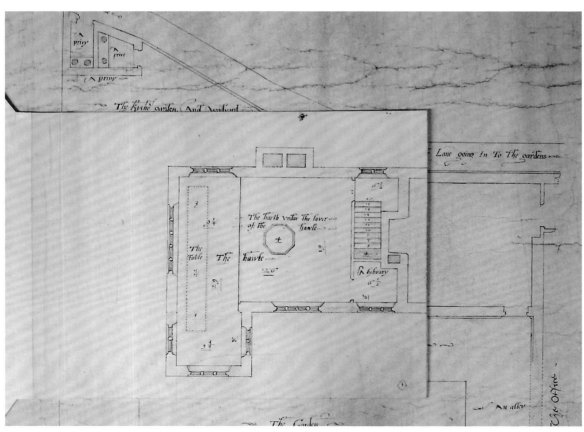

46, 47. John Symonds, plan of Cursitors' Hall with overlays, before 1579. National Archives, Kew. Catalogue 23

Compass and Rule

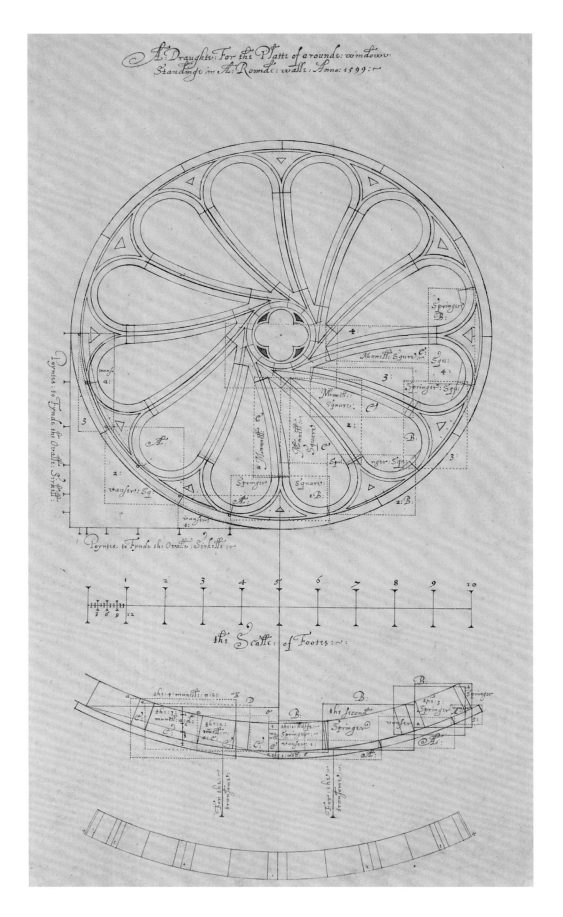

48. Robert Smythson, plan and elevation of 'a rounde window standinge in a rounde walle', 1599. Royal Institute of British Architects, London. Catalogue 24

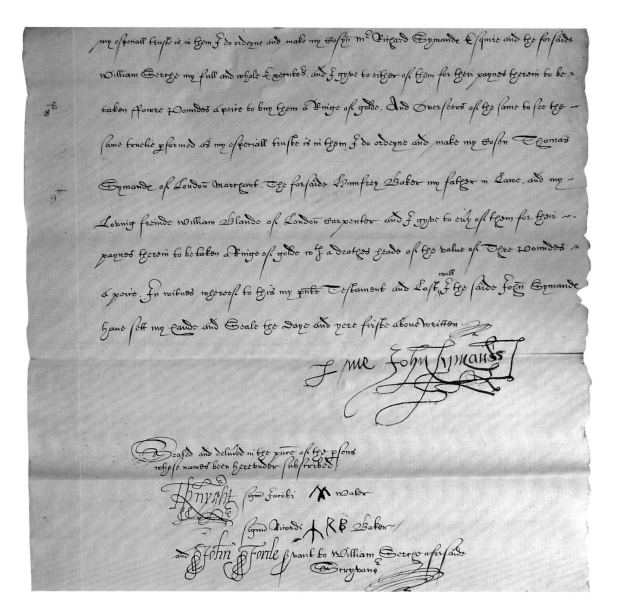

49. Will of John Symonds (last page, with signature), dated 1 June 1597. National Archives, Kew. Catalogue 25

the suitability of different types of local stone for the purpose.²² From his beginnings as a humble craftsman, he had remade himself into an almost universal technical expert.

Symonds's expertise in land, hydraulic, and quantity surveying extended, naturally enough, to cartography and architectural drawing. We include here two surviving instances of his graphic output. One notable result of his work at Dover, for example, is a plan of the harbour from 1583/84, signed and drawn to scale (figure 45, catalogue 22). Considering the size and difficult topography of the site, it was probably based on a measured survey. Symonds was a careful draughtsman and sophisticated in his preparation and use of plats. His smaller-scale work shows him using the paper medium to the full. A set of scaled plans of Cursitors' Hall provides a sheet showing the ground floor, with a cut flap to reveal the basement level below. The upper storey was also originally attached as a flap (figures 46, 47, catalogue 23). Flaps were used on harbour maps produced for Dover, and Symonds may have picked up this graphic technique there.²³ Their appearance in a context of civil building is unusual and suggests one way in which new ideas were transferred within the Elizabethan technical community. Indeed, P. D. A. Harvey has

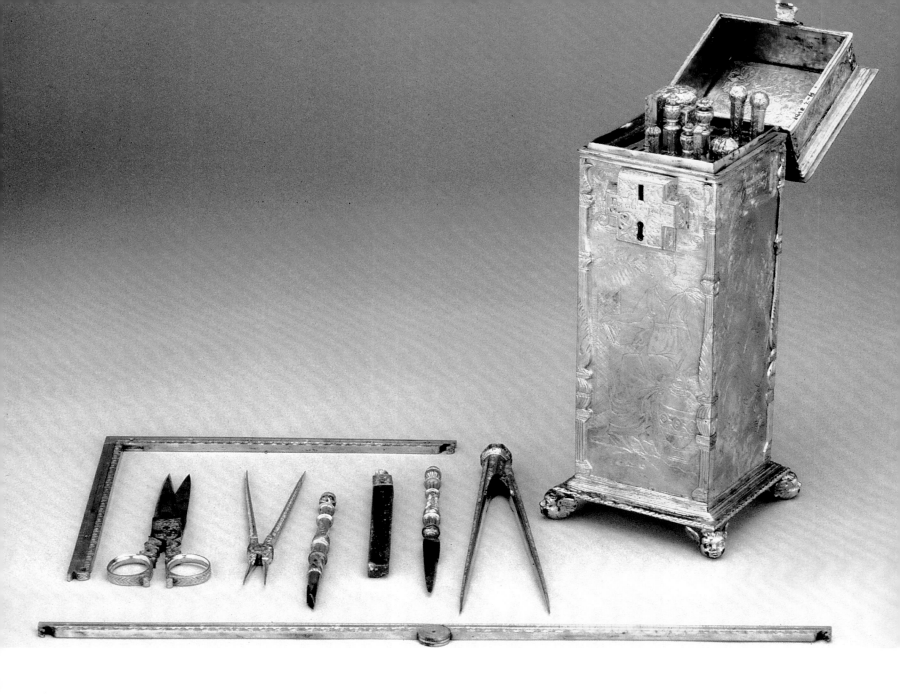

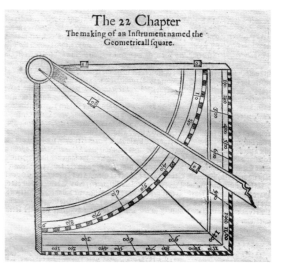

observed that Symonds's drawings for Cursitors' Hall are one of the earliest surviving sets of scale plans in England produced for a patron other than the Crown.[24] They are not dated, but must have been made sometime before 1579.

An analogous example of precision and originality in architectural drawing comes from the hand of Robert Smythson, a mason, surveyor, and architect. His plan and elevation of a twelve-light rose window 'standinge in a rounde walle' is dated 1599 (figure 48, catalogue 24). It is a stereotomic drawing, intended to provide step-by-step instructions for stone carvers to execute a very difficult and complicated form. In a straight wall, the component stones of the window can be made from identical moulds, but the curved surface gives them each a different inflection. The drawing reproduces an elaborate scheme for setting out the tracery, showing the masons how to use a single template for mirror-image pairs of voussoirs (*vausers*, marked A), springers (marked B), and mullions (*munnells*, marked C). The elevation gives the dimensions for the template faces, while the plan corresponds to those of the ends. Mark Girouard has observed that the drawing was probably connected with a round window of different design at Heath Grammar School, Halifax, built between 1597 and 1600. Similar distinctive rose windows would later appear in a number of great houses in the West Riding of Yorkshire. Given the drawing's unusual purpose, it seems likely that it reflects a tradition of full-scale setting-out that would normally have occurred on a tracing floor. Seen in this light, the precision and fastidiousness of the drawing are truly remarkable.[25]

The didactic and instrumental programme penetrated the traditional world of craft practitioners slowly and sporadically, but its consequences were nonetheless profound. Authors such as Recorde, Digges, and More offered an image of the mechanical arts ennobled by mathematics. Their view of the science was aspirational, in terms both social and intellectual. Mathematics, they argued, would raise craft traditions out of the realm of rote, uninformed experience, while at the same time enabling their practitioners to discourse equally with captains, statesmen, and governors.

John Symonds is an example of a master craftsman who achieved just such social and professional success. His will is richly revealing not only of his private life but also of his self-fashioning (figure 49, catalogue 25). One half of his estate was divided equally between his seven children, while the other half was distributed as a large number of specific legacies. Carefully enumerated and described, these bequests went to those in closest relationship to him, including both his relatives and his colleagues in the Office of Works. What stands out about the bequests is the way in which they intermingle conspicuous status markers, such as weapons and suits of rich clothing, with the trappings of Symonds's trade. John Allen, a favoured former apprentice, received 'my sky collored gowne and my rugge gowne … my backe sword and my second dagger my damaske Jackett with sleeves garded with velvet and my sylke grograyne doblett beinge cutt wth longe cuttes'. Such items would probably not have been out of place in the will of a wealthy merchant or lawyer, but, in addition to clothing, Symonds also gave Allen, 'halfe of all my tooles belonginge to my occupation as well for free stone as for hard stone and thone halfe of all my plats my best case of yron compasses with all the other tooles in the same case', as well as 'my finest pensell guilded which was my Mr. Stockettes'.[26] This last instrument obviously carried symbolic and sentimental weight. Symonds was passing it down to his apprentice just as his own master had done. The British Museum preserves the sole surviving example of an Elizabethan drawing set (figure 50, catalogue 26). Its components must be roughly comparable in quality to Symonds's 'pensell guilded', as well as to the instruments in his best case of brass compasses, which went to the Comptroller of the Works.

52 Humfrey Cole, altazimuth theodolite, 1586. Museum of the History of Science, Oxford. Catalogue 27

50. Bartholomew Newsam, set of drawing instruments, c.1570. British Museum. Catalogue 26

51. Geometrical square, from Digges, *Pantometria* (1571)

53. Humfrey Cole, simple theodolite, 1574. National Maritime Museum, Greenwich, NAV1448

The will also suggests something of Symonds's connection with the programme of mathematical practice. In addition to clothes, he bequeathed to one of his brothers 'my Geometricall instrument of wood called Jacobs staff', and to John Allen, 'my Geometricall square of latten for measureinge of lande'. The former instrument was a surveying cross-staff, similar to the device appearing on the title page of *Tectonicon* (figure 33). While this item is suggestive of Symonds's practical skills, it is the second instrument that arrests attention. 'Latten' is brass, which surely puts Symonds among the customers of a maker such as Humfrey Cole. The 'Geometricall square' could refer to either of two instruments that had recently appeared in Leonard Digges's posthumously published *Pantometria* (1571). One was mounted for horizontal measurement with a pivoted alidade or sighting rule at one corner (figure 51). The other was a component of a more exotic 'topographical instrument', an early form of altazimuth theodolite for measuring both horizontal and vertical angles. Cole made these universal surveying instruments, with some slight modifications (figure 52, catalogue 27). The inner square on the horizontal base is dignified as the 'Quadratum Geometricum' but appears on another instrument in the vernacular as 'geometrical square' (figure 53). Symonds may therefore have owned a simple theodolite of this latter type, in which the quadruple labelling of a part has become identified as the whole. Whatever the exact form of Symonds's instrument, he had clearly chosen a device from the newly developing repertoire of practical mathematics. The repeated use of the qualification 'Geometricall' suggests that these were not routine but distinctive and possibly prized objects.

The Mathematical Practitioner and the Elizabethan Architect

We do not know how enthusiastically he endorsed mathematics as a key to his own working identity, but his will provides strong evidence of its impact and personal significance.

Symonds's legacies provide a rich picture of his wealth and position, which he clearly sought to represent in his personal appearance. We find the same expression of status combined with craft pride manifested in several portraits of the period. William Portington, for example, was the King's Master Carpenter for fifty years, from 1579 to 1629. Toward the end of his life, he sat for a portrait commissioned by Mathew Bankes, one of the carpenters in the Office of Works and future Master of the Company (figure 54). Portington carries the serious and respectable mien of a royal official, delicately applying a pair of dividers to a rule.[27] A portrait of the shipwright Phineas Pett, of *c.*1612, makes an even more assured statement. The artificer is shown in the extravagant trappings of a courtier – a mark, no doubt, of the personal favour shown to him by Prince Henry and James I (figure 55). He also pretends to intellectual status. In his left hand, Pett brandishes a mathematical instrument: a sector of the kind only recently invented by Edmund Gunter. More sophisticated than a simple carpenter's folding rule, it allowed calculations involving square and cubic proportions and carried various trigonometric scales.[28]

Both Portington and Pett used instruments in their portraits, not only to identify their craft but to celebrate it in a broader claim to social status. Much the same strategy appears in a portrait at Emmanuel College, Cambridge of the successful mason-architect Ralph Simons or

54. Emmanuel Decritz (attr.), portrait of Master Carpenter William Portington, 1628. Carpenters' Hall, London

55. Portrait of Phineas Pett, with Gunter sector, *c.*1612. National Portrait Gallery, London

56. Portrait of Ralph Simons or Symons, early seventeenth century (?). Emmanuel College, Cambridge. Catalogue 28

Symons (figure 56, catalogue 28). The sitter – no relation to John Symonds – was one of the university's most prolific builders. He had worked from 1584 to 1587 at Emmanuel, in the years immediately after its foundation, adapting the ruinous Dominican priory buildings on the site for the college's use. In 1593, he was taken on by Trinity for a re-planning project, which eventuated in the Great Court. He was working there again, on the hall and kitchen, in 1604, and probably designed Nevile's Court in 1612. In 1596–98, he was building the newly founded Sidney Sussex, and in 1598 designed the second court of St John's. He was not an architect in the modern sense. Although he had a hand in designing all of these projects, his primary responsibility was in contracting and overseeing the work. Yet he was evidently held in some esteem. The portrait occurs in inventories of the College Gallery from 1719 and was probably made to hang, as it does now, among those of Emmanuel's notable fellows. A lengthy Latin inscription describes Simons's extensive work in Cambridge and identifies him self-consciously as 'architecti sua aetate peritissimi' ('the most skilled architect of his age'). Just as importantly, he is depicted wearing ruffed finery of a quality that John Symonds would have recognized, and displays a highly wrought mason's compass in his raised hand.[29]

Chapter Four
The Vitruvian Model: Inigo Jones and the Culture of the Book

The visual culture of the Italian Renaissance penetrated into England slowly and in piecemeal fashion. As a result, the public identity of the artist and the value ascribed to works of art remained largely dictated by medieval craft traditions well into the early modern period. The first architect to shake off these constraints was, famously, Inigo Jones, who self-consciously shaped his work and professional persona in an Italian humanist mould. Jones had journeyed to Italy twice for extended periods of observation and study. He became fluent in the language and deeply versed in the concepts and terminology of Italian artists and theorists. He was the first English architect to understand classical design not as a catalogue of decorative forms but as a complete and integrated system, one based on the reconstruction of ancient buildings and close observation of the work of modern masters, Andrea Palladio's in particular. As Surveyor of the King's Works, he brought this sensibility to the royal building administration, which he headed for over twenty years under both James I and Charles I.

Jones's sojourns were undoubtedly important to his education and training, but his approach to practice was equally shaped by Italian treatises. As an architect, he adhered to the entire repertoire of Vitruvian theory, fashioning a public identity almost entirely from texts. Foremost among his guiding principles was the distinction between theory and practice, and the supremacy of the former. For Jones, architecture was primarily intellectual in origin; craftsmanship and manual skill came second. The source of the work of art was the design, which originated in the mind and which called on the knowledge of all the arts and of many disparate disciplines. More than a means of material expression, architecture was a liberal art, a *scientia*, a body of knowledge.[1]

Jones defined this dichotomy in terms imported from Italian art theory and established it as the new basis of the builder's art. Consider, for example, the stage set for Jones's 1632 masque, *Albion's Triumph*, as one sign of his distance from craft tradition. The allegorical figures placed at the foot of the proscenium arch represented *Theorica* and *Practica* (figures 57, 58). As D. J. Gordon has pointed out, such figures were standard motifs on the title pages of architectural treatises and in this context served to underline the thematic importance of the play's architectural settings. Following Cesare Ripa's *Iconologia*, both figures are identified with the compass. The instrument associated with 'theory' is not held in her hand but rather opens upward from her mind, signifying the intellect's capacity to discover the measure and proportion of things. The attributes of 'practice', on the other hand, are the mason's yardstick and great compass. It is surely no accident that the figure is shown applying the tools to the ground, much as a mason would scratch out the form of a template on a tracing floor. What for medieval builders had been symbols of ancient authority, Jones transformed into emblems of rule, routine, and habit.[2]

We know about Jones's reading practice because a significant portion of his library has survived. Worcester College, Oxford owns some forty-six of Jones's own books, many laden

Compass and Rule

66

with annotations in his own hand. These texts and a handful of treatises held elsewhere allow us to follow the progress of his – largely self-guided – education. Judging by the extent of his marginal notes, the books that he used most often were Daniele Barbaro's edition of Vitruvius and Palladio's *Quattro libri*. The notes are roughly datable and they show him returning repeatedly to the texts – especially the latter – over a thirty-year period. The earliest annotations simply translate passages of interest, focusing on practical advice for building, the proportional intricacies of the orders, and the status of the architect. As his confidence grew, he began to grapple with the text, inserting criticism, comparing authors against each other, and reflecting on his own experience. This shift can be traced to the period of his second trip to Italy in 1613–14. He appears to have taken the book as a guide; several pages of notes show him comparing Palladio's plates with the buildings themselves. The same book also served as a practical reference for use in specific projects. John Newman has pointed to annotations made in 1619, as Jones sought solutions for some of the design problems of the Banqueting House at Whitehall.[3]

Jones was exceptional for the depth of his Italian culture, but the architectural treatise also enjoyed a certain influence among his contemporaries. Such publications were, after all, virtually the sole means by which Renaissance design principles and the professional Vitruvian ideal could be cultivated in England. We will also examine here the role of the Italian architectural treatise in two very different early seventeenth-century contexts: the academic milieu of the University of Oxford and the world of the surveyor and architect, John Thorpe.

The Vitruvianism of the continent overlapped with the medieval crafts in one important respect: both understood the design process in terms of a geometrical method. From here, however, they diverged. In the first place, practitioners of architecture *al'antica* prioritized – typically, though not exclusively – metrical and modular proportional schema. This was especially true for the orders, which most treatise writers defined in terms of multiples or fractions of a single module, usually set as the width or half-width of the column base. In this and in other respects, Palladio was Jones's guide and inspiration, but Vignola's method was among the most popular, as it assigned simple and easily remembered whole-number divisions to most parts of the orders. Jones himself seems to have recognized its ease of use and recall. As Gordon Higgott has observed, he employed Vignola's twelve-part modular division on at least one occasion: it appears on a design for a gateway at Arundel House in the Strand in 1618.[4]

Although less cherished than his Palladio, Jones's copy of Vignola's *Regola delli cinque ordini* also bears evidence of a close and attentive reading. The annotations – many in Italian – show Jones interacting with the text on a number of levels. Brief and telegraphic, they served alternately to record the ancient sources of Vignola's 'inventions' (pls. 13, 19) or to note the architect's deviation from other authorities and from ancient practice (pls. 12, 26). Many annotations show a keen awareness of minute proportional relationships between different members and the effects produced by altering them. These remarks drew attention, for example, to the projection of engaged columns in Vignola's Doric arcade (pl. 10) or the depth and width of channelling in a rusticated wall (pl. 32). Jones took a particular interest in Vignola's Attic base, which differed considerably from that of Palladio, in that it had a deeply undercut scotia, the inward curving moulding between the upper and lower torus (figure 59, catalogue 29). Vignola had provided a partly modular and partly geometrical method for drawing this curve, which was composed of two circles of different radius. Jones's annotation shows him not only deciphering the construction but also trying to simplify it. The attention that Jones paid to this detail suggests not only that the form intrigued him but that he was trying to find a way of replicating the construction at full scale.[5]

57. Inigo Jones, allegorical figure of 'Theory', 1632, detail from a drawing of a proscenium arch for the masque *Albion's Triumph*. Devonshire Collection, Chatsworth

58. 'Practice' from *Albion's Triumph*

59. Page from Inigo Jones's annotated copy of Vignola, *Regola delli cinque ordini* (1607), showing diagram for the construction of the Attic base. Worcester College, Oxford. Catalogue 29

The Vitruvian tradition departed from craft practice in a second respect. In classical design, 'geometry' was identified exclusively with the conception of the work and with the preliminary process of planning and composition. In contrast to the sometimes piecemeal process of Gothic building, this approach demanded a greater attention to the relationships between the parts of a building and the work as a coherent whole. Proportional design methods obtained here too, with ratios defined either in terms of a common module or – more frequently – as a correspondence between the dimensions of individual elements and spaces. The Vitruvian term *symmetria* summarized both the principle involved and the effects produced: clarity, coherence, and consistency. In a famous article, Rudolf Wittkower illustrated how Jones actually carried

The Vitruvian Model: Inigo Jones and the Culture of the Book

out this strategy in the context of a specific commission, namely his remodelled façade for the medieval cathedral of St Paul in 1633–34 (figure 60). Underneath the inked drawing lay a net of scored lines and compass pricks, which Jones used to set out the principal elements of the façade. As Wittkower showed, Jones divided the far left pilaster shaft into three equal parts. At the level of the first third, he set out a horizontal scoreline that served to coordinate several other elements, including the height of the aisle doors, the width of the outer bays, and the column shafts framing the central door. The diameter of the lower order was further subdivided to provide measures for setting out the parts of the entablature and the width of the second-storey columns.[6]

60. Inigo Jones, proposal for the façade of St Paul's cathedral, 1633–34. Royal Institute of British Architects

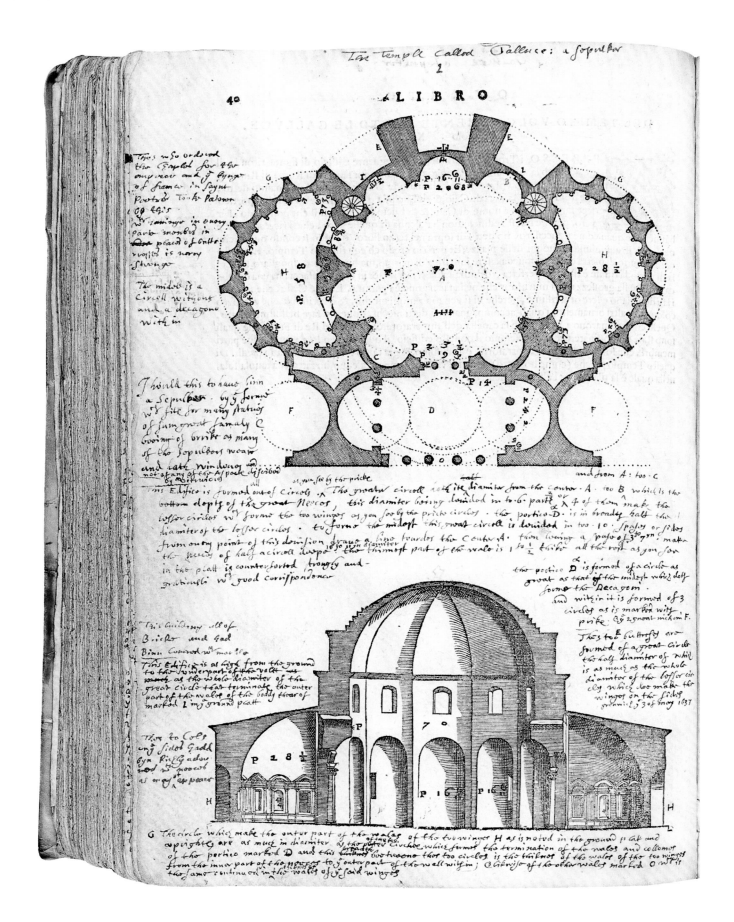

The Vitruvian Model: Inigo Jones and the Culture of the Book

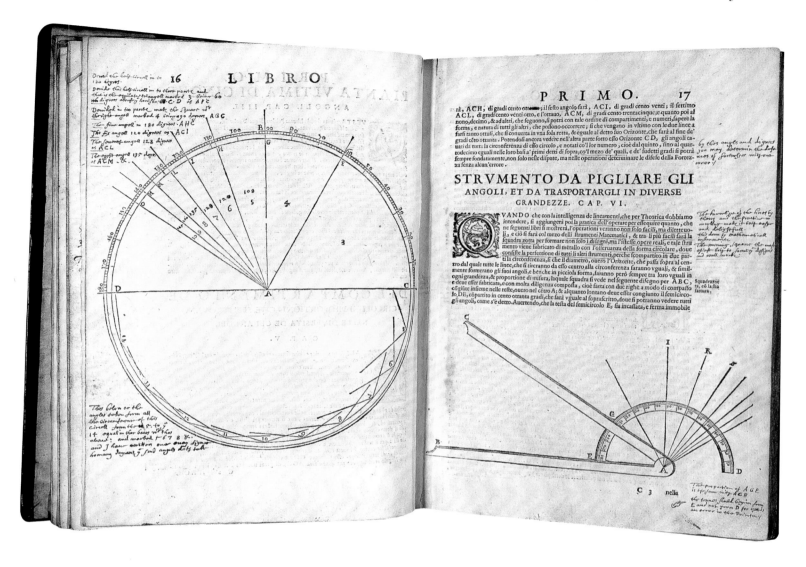

One of the plates of the *Quattro libri* appealed to Jones as a perfect illustration of this principle. The fourth book (p. 40) contains a plan and section of the domed temple of Minerva Medica in Rome, referred to by Palladio as 'Le Galluce' (figure 61, catalogue 30). The page is covered with notes, the longest of which, dated 'grenwich ye 3 of May 1637', concerns the composition of the plan, formed from concentric and interlocking circles. Jones was evidently attracted not only to its complexity but also to its internal harmonies. He used his dividers to seek out the principal ratios and then pricked the scored arcs with his pen, to create dotted outlines. As he discovered, the width of the interior space of the dome served to define both the exterior envelopes of the wings and the depth of the portico. Moreover, the inner diameter of the wings corresponded to two-fifths the breadth of the interior and also provided the radius for the external buttresses of the dome, to the rear of the building. The length of the portico was three times its width. Jones tried to find a module that linked the different parts of the plan to a common measure, but with limited success. In his notes, he suggested that the inner diameter of the wings corresponded to the interior width of the dome measured from opposite niche-ends (C to B on the plan) in a ratio of two to three. His calculation appears to be correct, but it is notable that the unit thus implied was used to generate few of the other dimensions. This circumstance, however, did nothing

62. Annotated pages showing mathematical instruments, from Jones's copy of Lorini, *Le fortificationi* (1609). Worcester College, Oxford. Catalogue 31

61. Temple of 'Le Galluce,' printed plan with inscriptions from Jones's copy of Palladio, *I quattro libri* (1601). Worcester College, Oxford. Catalogue 30

to lessen the architect's enthusiasm. Even without a systematic reference to a common module, the design was, as Jones declared, 'countersorted strongly and gratiousli with good correspondence'.[7]

Military architecture offered an analogous opportunity for geometrical design, and it is perhaps not surprising to find Jones drawn towards that subject. Vitruvius, after all, had defined fortification as an integral part of architecture. Like many architects of the period, Jones would have been able to see it as a component of both classical building tradition and the modern universe of mathematical arts and sciences. Of course, Jones had little occasion to practise the art. He received no such commission that we know of as Surveyor of the King's Works, but that did not prevent him from tackling the subject with his typical diligence. Among the several books from his library on military art and history, one of the most extensively annotated is Buonaiuto Lorini's *Le fortificationi*. In addition to the detailed recommendations on the layout of bastions (pp. 20–21), Jones was particularly interested in the specialized instruments that Lorini had illustrated (figure 62, catalogue 31). These included a graduated disc to help form polygons and an angle-measuring instrument equipped with sighting vanes. The latter was to be used for transferring degree measures from a paper plan to the worksite and vice versa. In describing the device, Lorini articulated one of the mathematical practitioner's central tenets, which Jones paraphrased in the margin of the text: 'The knowlige of the lines by theory and the practice in working make it both easye and delightfull, this done by mathematicall instruments'.[8]

Jones's manner of reading was attuned to his work as a practising architect and to the example set by the Italian *trattatisti*, Palladio in particular. It is worth underlining, however, Jones's exceptional status. Aside from his pupil, John Webb, few of Jones's contemporaries were able or inclined to immerse themselves in the subject with the same level of detail and commitment. Other readers used the architectural treatise in different ways. The next two sections illustrate contrasting instances of the reception of Renaissance architecture in England. As we will show, the art's geometrical content – in both practical and symbolic terms – was central to both.

In 1620, Clement Edmondes, a diplomat and municipal official of London, donated a 'mathematical model' to Thomas Bodley's newly established library (figure 63, catalogue 32). This alabaster sculpture, standing some twenty-eight inches high, consists of a pentagonal pillar, crowned with a skeleton dodecahedron. Within this shape hangs a more involved, complicated form: a brass 'pentakis' dodecahedron, fixed in place by a metal rod. Around the base of the pillar are arranged six additional geometric solids, the remains of what was once a larger complement. Four of these are of brass: a cube, a tetrahedron, an icosahedron, and a dodecahedron. The remaining two, an icosahedron and a fragment of an octahedron, are carved from alabaster. In addition to the regular or 'Platonic' solids, the sculpture evokes a second theme: the classical orders. The faces of the central post are decorated with rusticated pilasters corresponding to the five orders of architecture. Each pilaster stands on a high pedestal and is surmounted by an articulated Composite entablature. The object is nothing if not enigmatic. Although the two themes are evidently related via a common basis in geometry, in many ways the model seems expressly made to provoke speculation as to its purpose and meaning. Fortunately, Alexander Marr has recently devoted an in-depth study to the work and its donor. Our account attempts to build on his very thorough interpretation.[9]

This unusual object was, in the first place, a gift. As Bodley worked to restore the Duke Humphrey Library as a permanent foundation for the University, he used his wealth and influence to cultivate a strategy of competitive patronage, 'to stirre up', as he put it, 'other men's

63. 'The Architectural and Mathematical Model' of Clement Edmondes, c. 1620. Museum of the History of Science, Oxford. Catalogue 32

Compass and Rule

64. Portrait of Henry Savile, unknown date. Museum of the History of Science, Oxford. Catalogue 33

65. 'Tower of the Five Orders', Bodleian Library, Oxford, from 1613

benevolence'. Not only did he encourage gifts of books, he also worked to incite potential donors to outdo each other, by making their gifts conspicuous. The names of donors and the size of their gifts were publicized in printed lists and recorded in a lavish register book. The policy was continued following Bodley's death in 1613 by Henry Savile, his friend and principal advisor in matters relating to the design and fitting of the library. It was Savile who supervised the eastern expansion of the library in the form of the Schools Quadrangle, still underway at the time of Edmondes's donation. Its entry in the Benefactor's Register still reflects the founder's enthusiasm for objects of wonder and curiosity, were they books and manuscripts or rare and unusual artefacts. The sculpture is described as an 'excellent *paradeigma* of five columns, now invented for the first time according to rustic form, having been made from alabaster with matchless skill'. The register records comparable gifts in the period between 1601 and 1636: an armillary sphere, mathematical and astronomical instruments, ancient coins, modern medals, even an Egyptian idol. Many of these artefacts were displayed in the University picture gallery, in what is now the upper reading room on the top floor of the library.[10]

Edmondes had been cultivating contacts at Oxford for many years. He knew Bodley well, having donated a similarly rare and exceptional object to the library in 1607 – a Hebrew cabbala in the form of an illuminated parchment roll. The donor must also have had some connection with Savile or was trying to create one, for the themes evoked by the model would have resonated deeply with his interests as a mathematician, teacher, and classical scholar

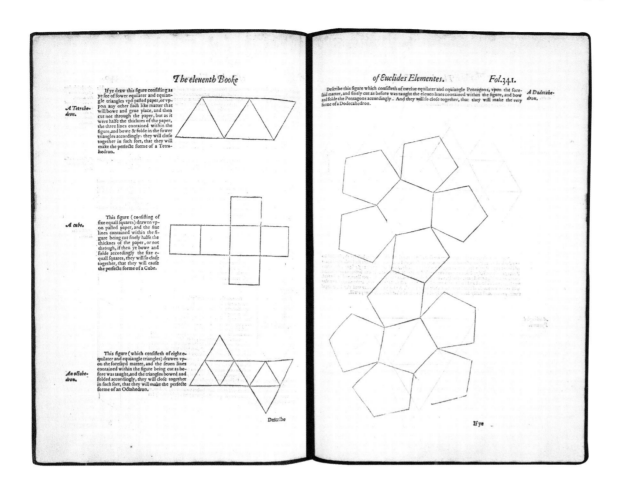

66. Cut-out solids from Euclid, *The Elements of Geometrie*, trans. H. Billingsley (1570). Bodleian Library, Oxford. Catalogue 34

(figure 64, catalogue 33). As commentators have long noted, the work may have been intended as a demonstration aid in the teaching of mathematics, perhaps for the newly endowed chairs of geometry and astronomy that Savile founded in 1619. As a visual aid, it might easily have been used for the culminating discussion in Book Thirteen of Euclid's *Elements*, on the properties of the five regular solids. Such models were not uncommon in teaching the subject: John Dee, for example, warned that the final books of the *Elements* were 'somewhat hard for young beginners, by reason they must, in the figures described in a plaine, imagine lines and superficies to be elevated and erected, the one to the other'. Some of these 'solides or bodies', Dee continued, 'which … [students] have not hitherto bene acquainted with, will at the first sight be somewhat straunge unto them'. The translator, Henry Billingsley, took this caution to heart, by including cut-out patterns and even pages with pasted flaps of paper, which readers could fold up to make their own three-dimensional forms (figure 66, catalogue 34).[11] Wooden and glass versions are often represented in a pedagogical context, as in the famous portrait of Luca Pacioli attributed to Jacopo de'Barbari (1495, Museo di Capodimonte, Naples) or Nicholas Neufchatel's likeness of Johannes Neudorfer and his son (1561, Alte Pinakothek, Munich).

The model was also related to Savile's interest in architecture and his activities as a patron. As Warden of Merton College, he had introduced classical design on a grand scale to Oxford, having commissioned the four-storey 'frontispiece of the orders' in the Fellows' Quadrangle in 1608–10. He was probably also responsible for the decision, sometime soon after 1613, to reuse the motif in the Bodleian's Tower of the Five Orders (figure 65). Such displays of superimposed columns had previously been limited to aristocratic town and country houses. As John Newman has observed, the Merton and Bodleian frontispieces introduced what was essentially a palatial and seigneurial motif into the academic world, thereby transforming its connotations. Another indication of Savile's input lies in the 'academic' nature of the composition. The Bodleian tower – even more elaborate than Merton's – rises through five storeys of paired fluted columns, each projecting from the façade and resting on the entablature of the order below. Beginning with the Tuscan and climbing up through the Doric, Ionic, and Corinthian to the crowning Composite, they diminish in scale as they rise, becoming more slender and elaborately ornamented. In its upward progression and diminution, the composition reflects a close attention to the rules for superimposed orders set out by Vitruvius and his Renaissance commentators, but it is notable that none of its elements are integrated into the façade as a whole and that its classicism is at variance with the Gothic used everywhere else in the library. The architecture of the tower is a decorative appliqué, used principally for its iconography – in this case, as a symbol of the nobility of learning.[12]

The mathematical model makes a similar statement, associating architecture – particularly the orders – with knowledge itself. Plato, in the *Timaeus*, had ascribed cosmological significance to the five regular bodies as symbols of the elements. He associated the cube with earth, the octahedron with air, the pyramid with fire, and the icosahedron with water. The dodecahedron, which most closely approaches the sphere, was associated with the cosmos, for this form 'God used as a model for the twelvefold division of the Zodiac'. This connection was still powerful in the early seventeenth century. Indeed, Johannes Kepler had recently reaffirmed it: his *Mysterium Cosmographicum* (1596) attempted to account for the gaps between adjacent planetary orbits by relating them to the inscribed and circumscribed spheres of the five regular polyhedra. Edmondes's model draws implicitly on the same cosmic symbolism: the central pillar rises up from the base – where it is surrounded by the elemental forms – to carry a large dodecahedron, carved from the same alabaster. Architecture here becomes a symbol of pedagogy, wherein the progress of knowledge via pure geometry reaches from the realm of the mundane into that of the abstract and eternal. At the same time, the model makes a second claim about the nature of architecture itself, as an art that participates simultaneously in both the material and the incorruptible world. This Vitruvian commonplace was particularly emphasized in Dee's 'Mathematicall Praeface' to Billingsley's Euclid. 'With … the chief Master or architect', Dee maintained, paraphrasing Leon Battista Alberti, 'remaineth the demonstrative reason and cause of the mechaniciens worke in lyne, plaine, and solid'.[13] The fact that the five Euclidian solids are matched by the same number of canonical orders also supplies part of the work's basic conceit.

The model appears to profess pedagogic value for the teaching of architecture, but it is doubtful that this idea was immediately implemented. Savile was certainly not unsympathetic to the practical mathematical arts: optics, gnomonics, geography, and navigation all fell within the province of the Savilian professorships. Architecture, however, was not included among these, nor is it clear how the model might have been used in such a context. It is possible that the rusticated segments on the faces of the pilasters – which differ in number from one to another – were intended as a mnemonic device for the modules of each order. The Tuscan pilaster has six, the Doric ten, the Ionic twelve, the Corinthian seven, and the Composite fourteen. These divisions, however, do not match those recommended by any of the major treatise writers and, in the absence of other sources, it is difficult to assign the model a more

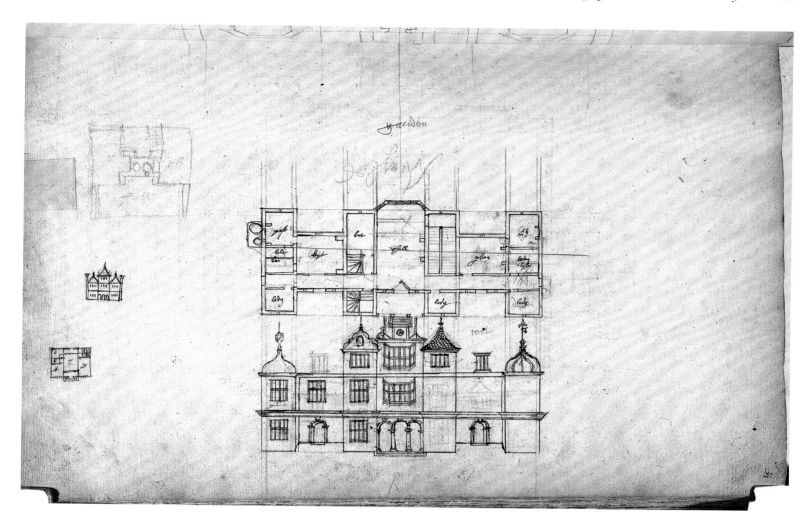

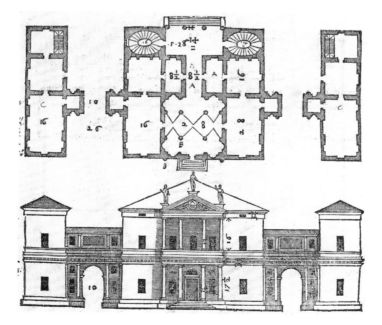

67. House plan and elevation from John Thorpe's album, Sir John Soane's Museum, T34

68. Plan and elevation of the Villa Pisani from Palladio, *I quattro libri* (1601)

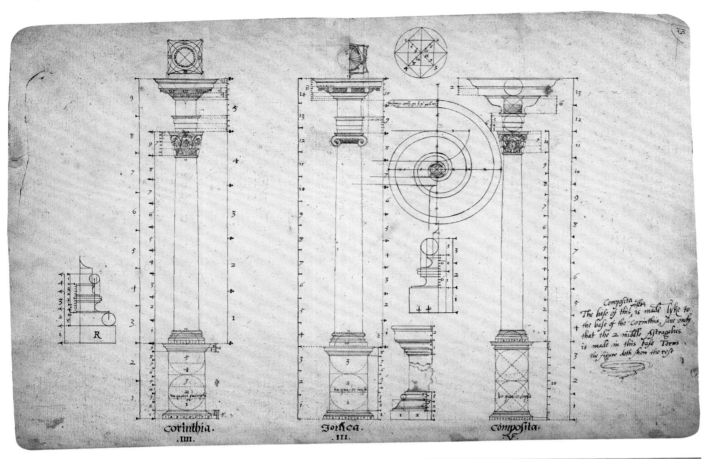
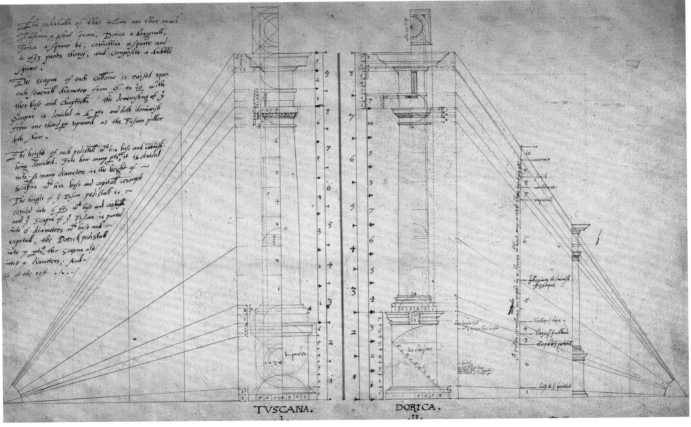

specific teaching role. Yet, its reliance on Italian architectural culture and the Vitruvian ideal is nonetheless evident. The novelty of the model, like that of the Bodleian's Tower of the Five Orders, lay in the way in which it deployed the classical orders as a symbol of the education of young gentry and noblemen. Such an understanding was naturally suited to a context of humanist learning, particularly in the course of establishing a library, for the image of architecture that it assumed was shaped by the printed book.[14]

Many of the principal sources for the use of architectural books in early seventeenth-century England refer us to readers like Savile. As members of a literate and wealthy minority, they were able to travel abroad and amass libraries that tended to reflect their habits and interests as patrons. Lucy Gent and Malcolm Airs have recorded a number of considerable sixteenth-century collections of this kind, notably those of Sir Francis Willoughby at Wollaton Hall; Sir Thomas Smith at Hill Hall; Henry Percy, Earl of Northumberland at Petworth and Syon; and the largest of the time, Sir Thomas Tresham's at Rushton and Lyveden. In contrast, we know far less about the more common readers of architectural books, namely craft practitioners. Although books are sometimes mentioned in craftsmen's wills, the volumes tend to be few in number and their titles are rarely itemized. Like Inigo Jones, they would have read primarily with practical intent, although without his broader intellectual concerns or his meticulousness. For most craftsmen, architectural treatises seem to have served as pattern books, as convenient visual resources. Their role was to supply ideas and motifs that could be borrowed and adapted according to need.[15]

It is this set of circumstances that makes John Thorpe such an unusual figure. This architect and surveyor came from a long line of Northamptonshire masons. His brother, Thomas, was also a member of the profession and, although John himself does not appear to have been trained in the family craft, he did not stray far from it. In 1583, he became a clerk of the Queen's Works, where he was employed as an accounts-keeper and draughtsman. We know that he carried out various small-scale improvements at Richmond, Greenwich, Whitehall, and many other royal palaces. He was therefore a colleague of John Symonds (discussed in Chapter 3) and was remembered along with other Works' personnel in the latter's will. In 1601, he left the Crown's service and established what would become a successful and long-lasting private practice as a surveyor of both estates and buildings. Despite his practical background, Thorpe was not unlettered. Although he did not match Inigo Jones's total immersion in Italian erudite and artistic culture, he was nevertheless able to pursue an interest in continental architecture that went beyond that of most craftsmen of the time. His contemporary, Henry Peacham, called him an 'excellent geometrician and surveiour … not only learned and ingenious himselfe, but a furtherer and favourer of all excellency whatsoever, of whom our age findeth too few'. In many ways, Thorpe represents the impact of the Italian Renaissance not in the rarefied environment of the court but in the everyday world of the guild-trained practitioner.[16]

Thorpe would no doubt have remained an obscure and largely forgotten figure were it not for his famous album of architectural drawings, rediscovered by Horace Walpole in 1780 and now preserved in Sir John Soane's Museum. This manuscript volume contains some 150 house plans and roughly 30 elevations. It is the kind of collection – like Robert Smythson's, now at RIBA – that must once have been commonly assembled over the course of a craftsman's life. Such 'plats' often feature in mason's wills, where they are treated as treasured heirlooms; as John Summerson has shown, Thorpe's own collection originated with thirteen plans inherited from his father in 1596. The drawings form two groups related to Thorpe's work as a

69. Drawing, with details and notes, of the Corinthian, Ionic, and Composite orders, from John Thorpe's album. Sir John Soane's Museum, London. Catalogue 35

70. Tuscan and Doric orders, John Thorpe's album

Compass and Rule

71. Title page, from Blum, *The Booke of Five Collumnes* (1601). Worcester College, Oxford. Catalogue 36

surveyor. The larger contains designs of his own invention, some of which were even built. He appears to have provided plans to prospective builders, although he did not generally supervise their construction. The balance of the album represents pre-existing buildings that Thorpe had either measured himself or that he reproduced from drawings.

Interspersed among these two groups are several drawings of foreign derivation, and it is from these that we can glean something of Thorpe's response to European architecture. Most, not surprisingly, are copied from printed works. Judging from the visual references, his favoured authors were Palladio, Vredeman de Vries, Hans Blum, and especially Jacques Androuet Du Cerçeau. Thorpe approached these sources in a way that must have been typical of contemporary architects, absorbing ideas and 'anglicizing' them according to current Elizabethan and Jacobean conventions. Summerson noted one telling instance involving a direct quotation from Palladio's Villa Pisano taken from the engraving in the *Quattro libri* (figures 67, 68).[17]

For a practitioner, Thorpe had unusually high literary ambitions, and the album contains unique evidence of these. Among the volume's most polished drawings are two fine studies of the five orders, copied from Hans Blum's *Quinque columnarum exacta descriptio atque deliniatio* (1550) (figures 69, 70, catalogue 35).[18] The content and style of these drawings allowed Summerson to identify Thorpe as Blum's English translator, named only as 'I.T.' in *The Booke of Five Collumnes of Architecture* (1601). Indeed, Thorpe alludes to himself with the same initials in several of the handwritten inscriptions in his album. This set of studies does not appear to have been made merely as an aide-mémoire. The draughtsmanship is detailed, exact, and highly polished – far more so than virtually any other drawings in the album. It is possible that they were initially intended for publication. If that is the case, the proportioning instructions inscribed at the sides of the drawing may represent directions for the engraver.

The value of Blum's little book lay in the simplicity and flexibility of its proportional rules, which the author had adapted from Serlio's modular system into an easily understood graphic form. In addition to (and separate from) this system, Blum also introduced a new proportional device of his own invention: a division of the entire order including pedestal and entablature into equal parts. This second scale enabled practitioners to proceed from any chosen height to the correct proportioning of the whole order, something that Serlio's module did not allow. Both of these systems are explained graphically in Thorpe's drawing of the Tuscan and Doric (T 14). The vertical scales running along both sides of the drawing's centre-line represent Blum's equal-parts division, whereas Serlio's modular scheme is set out on the opposite sides of the

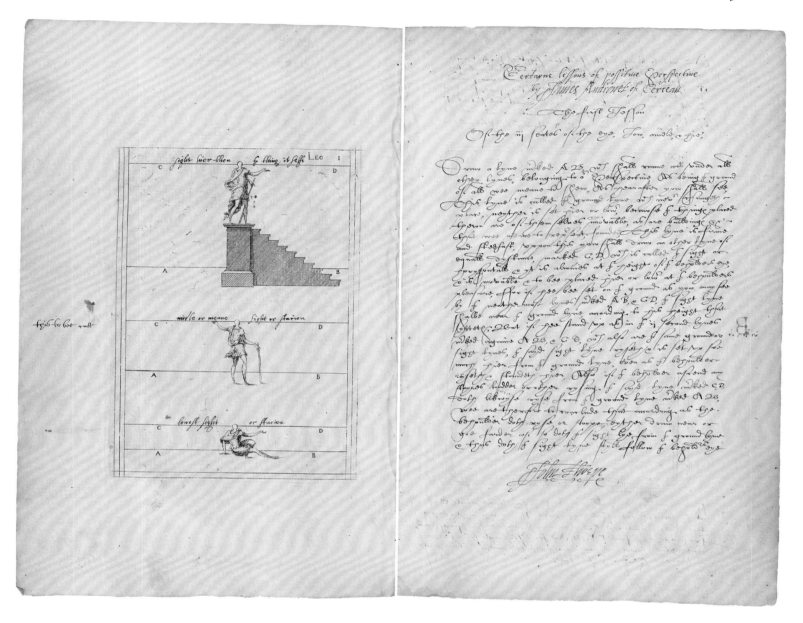

two orders. In contrast to Blum's scales, the modular arrangement requires a separate method of proportional reduction, hence the lines converging to points toward the edges of the page.

Thorpe's decision to translate this little book was based not on the fame or authority of the author – Blum was no Palladio – but rather on the work's utility. It was intended, as the title explained, 'for the benefit of Free-Masons, Carpenters, Goldsmithes, painters, Carvers, In-layers, Anticke-Cutters, and all other that delight to practice with the Compasse and Squire'. In this respect, Thorpe was prescient. His translation became one of the most successful books on the orders in England, appearing in seven editions by 1678. The sole surviving copy of the first edition forms part of George Clarke's personal collection, which he left, along with Inigo Jones's volumes, to Worcester College (figure 71, catalogue 36).[19]

There exists another remnant of Thorpe's literary career: a manuscript translation of Jacques Androuet Du Cerceau's *Leçons de perspective positive* (1576) held in the Bodleian library. Like Blum's little book on the orders, Du Cerceau's is a brief, practical tract designed explicitly for

72. 'The First Lesson' from John Thorpe's manuscript translation of Du Cerceau, *Leçons de perspective positive* (1576). Bodleian Library, Oxford. Catalogue 37

beginners. It is organized into discrete lessons, systematically setting out the principles of the technique. Of the book's sixty chapters, the first nine are in Thorpe's hand, with the remainder completed by a professional scribe (figure 72, catalogue 37). As Karl Josef Höltgen has observed, the manuscript is evidently a fair copy intended for publication and was probably composed *c.*1603, making it the earliest perspective manual in English. Although he was very much embedded within an English craft tradition, Thorpe's interest in writing and publishing nevertheless marks his orientation towards a novel – and foreign – professional ideal.[20]

Chapter Five
Vision, Modelling, Drawing: Christopher Wren's Early Career

Inigo Jones's vision of the art – rigorous, learned, and idealizing – was essentially alien. Despite the profound influence that his architecture would later exert, he could provide no professional exemplar for contemporary craftsmen. For most of the seventeenth century, the humanist conception of architectural design remained the province of gentlemen *virtuosi*, who saw the art as a complement to other forms of polite learning. Figures such as Hugh May or Roger Pratt were initially attracted to architecture as cultivated and travelled amateurs. Although they practised in a sustained and devoted fashion, they did so with no formal claim to the title of architect, but typically as an advisor or administrator in the service of a nobleman or the Crown. Pratt's well-known advice to the prospective house-builder in search of a plan is often seen to describe himself:

> Get some ingenious gentleman who has seen much of that kind abroad and been somewhat versed in the best authors of Architecture: viz. Palladio, Scamozzi, Serlio etc. to do it for you, and to give you a design of it on paper, though but roughly drawn, (which will generally fall out better than one which shall be given you by a home-bred Architect for want of his better experience, as is daily seen).[1]

Note that the author conspicuously avoids identifying the 'gentleman' as an 'architect', applying the latter term instead to the 'home-bred' builder. For Pratt, it was important to define his involvement in the art not as a profession – even one yet to be born – but as a disinterested *avocation*. As Howard Colvin has observed, such men did not, as a rule, involve themselves in contracting, nor were they generally paid a fee for their services, preferring rather to receive favours, gifts, and offices.[2]

Pratt's conception of architectural practice – and of his own role as a designer and overseer – owed much to the Italian–Vitruvian tradition. His time in Rome, his travels on the continent, and the treatises he mentions suggest as much. This view of the art was complemented, however, by a deeply rooted, native history of mathematical practice. Indeed, by this time, both traditions had coalesced within a broader ideal of virtuosity. John Webb, for example, could confidently equate design with the 'arts mathematical' and defend his former master as a 'great Geometrician'. John Evelyn likewise expanded on the consummate knowledge required for 'the flower and crown as it were of the sciences mathematical'.[3] Roger North, another gentleman-practitioner, described his interest in architecture as a natural outgrowth of 'these joys I had in mathematicall exercises', embracing perspective, optics, and instrument-making. The rebuilding of the Middle Temple, following the fire of 1679, occasioned his introduction to the art: 'I had the drawing the model of my little chamber, and making patterns for the wainscot, and from thence the practice of working from a scale, all the while exercising the little practical geometry I had learnt before, and in short found the joys of designing and executing known only to such as practise or have practised it'. In his autobiography and his unpublished architectural writings, North endorsed architecture in the same terms that had long been employed in English mathematical apologetics: it was a source of keen intellectual pleasure and a practical activity conducive to the public good.[4]

This context informs the career of the most celebrated 'amateur' architect of the time: Christopher Wren. Like the engineers of the sixteenth century, Wren saw architecture in terms of a global vision of mathematics. Indeed, his interest in the art is almost inconceivable without it. For Wren, however, the whole was additionally shaped by a culture of mechanical invention, by antiquarian study, and by recent advances in physical and structural mechanics. Nor was this view of the art unique. We find the subject explored under the same range of headings in the meetings of the Royal Society, following its foundation in 1660. These discussions touched on building materials and techniques related to the group's interest in the history of trades, on archaeological remains and reconstructions, on new surveying and drawing instruments, and on problems involving the resistance of beams and the efficiency of arches and domes. Design issues as such were rarely discussed but, in most other ways, architecture provided a suitably diverse subject for the early Society's informal and still-unstructured programme.[5]

We begin in Wren's youth, with a question that has long bedevilled historians. Why architecture? Wren had become a fellow of All Souls in 1653, professor of astronomy at Gresham College in 1657, a founding member of the Royal Society in 1660, and Savilian professor of astronomy at Oxford in 1661. By this time, he was already widely celebrated as an astronomer and practical mathematician. William Oughtred, the doyen of English mathematics, had called attention to his talent as early as 1652. The teenaged Wren, he enthused, had 'enriched astronomy, gnomonics, statics and mechanics with brilliant inventions, and from that time has continued to enrich them, and in truth is one from whom I can, not vainly, expect great things'.[6] Two years later, John Evelyn referred to him in his diary as 'that miracle of a youth' and 'that prodigious young scholar', and in 1662 offered him a more public encomium as 'that rare and early prodigy of universal science'.[7] Isaac Barrow, in his inaugural address at Gresham the same year, praised Wren's 'divine genius'. 'Once a prodigy of a boy', Barrow proclaimed, 'now a marvel of a man … his merit attracts the eyes of the whole world.'[8] These witnesses were not referring to Wren's abilities as an architect, for, indeed, he had yet to express any interest in the art. Nor did his background or recent achievements suggest that he intended to do so. If anything, his elevation in 1669 to the office of Surveyor-General of the King's Works provides even greater cause for surprise, for the extraordinary duties of the post following the Great Fire of 1666 effectively curtailed his full engagement with science. What connection did Wren see between architecture and the mathematical sciences that allowed him to move from one to the other? Should this shift be explained merely as a change in taste, a decision to pursue more worldly ambitions, or did it rely on and represent some structural affinity between the two fields?

Wren's youthful exercises and inventions are difficult to categorize in strict disciplinary terms. In general, they fall under the broad umbrella of the mathematical sciences, astronomy in particular. One of his earliest known letters, written to his father when he was only thirteen years old, mentions an 'astronomical instrument' in the form of a star calendar, a 'pneumatic engine' or air pump, an 'exercise in physics' on the origin of rivers, and another on optics.[9] A letter of 1647–48 relates his invention of a recording 'weather clock' and a brass disc on which was

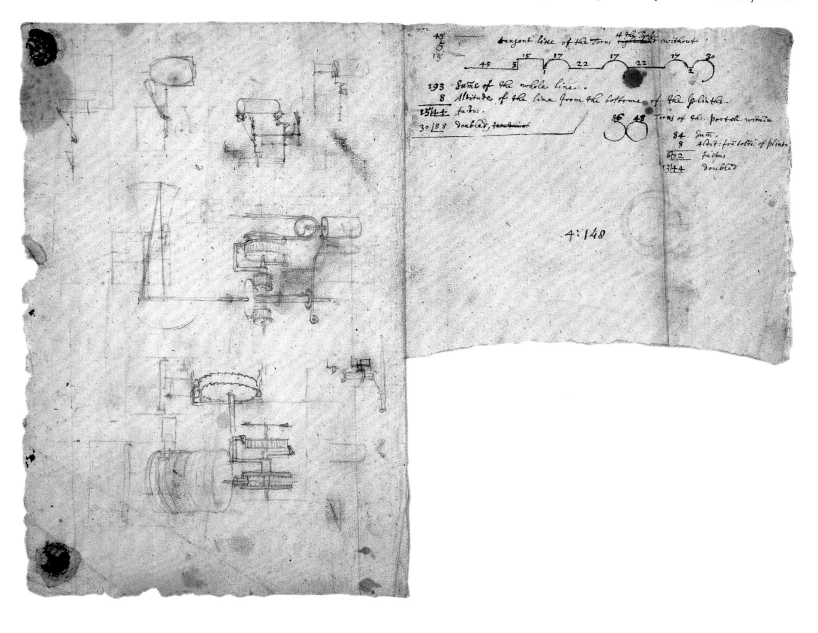

engraved 'an epitome' of 'the whole theory of Spherical Trigonometry'. Both of these objects were devised with the help of the physician and anatomist Charles Scarburgh, with whom Wren lived for some time before matriculating at Oxford.[10] The Wren family *Parentalia* describes Christopher as Scarburgh's assistant in his dissections and anatomical experiments, even constructing pasteboard models of muscles for the older man's use.[11] Wren's sundials are mentioned in a letter from Thomas Aylesbury in 1649, including one designed for his father's house, the reflected face of which was projected onto the walls and ceiling of a room.[12] Works from 1650–52 include a 'double writing' instrument, an algebraic treatise on the Julian period, and a Latin translation of Oughtred's tract on dialling.[13] In 1653, Wren invented a perspective machine, and in 1655 was reportedly preparing a book of microscopic observations, as well as a topographical lunar globe and an account of the moon's librations.[14] As many historians have

73. Christopher Wren, studies for a weather clock, c.1663–64. All Souls College, Oxford. Catalogue 38

observed, these interests were cultivated by John Wilkins, Warden of Wadham College. In addition to Scarburgh, Wilkins's circle included the mathematicians Seth Ward and John Wallis, the experimenters Robert Boyle and Robert Hooke, the physicians William Petty and Thomas Willis, and the astronomer Lawrence Rooke.[15]

The character of Wren's early enthusiasms marked virtually his whole career. His later work, for example, on the description of Saturn, the rectification of the cycloid, and the physics of impact emerged from the same set of investigative practices and techniques that he had developed in boyhood. Even as he turned toward more advanced work in astronomy, pure mathematics, and natural philosophy, Wren continued to think in terms of new and ever more creative forms of instrumentation, visual display, and ingenious contrivance. Indeed, he persisted with some of his youthful ideas and inventions – the weather clock and the perspectograph in particular – for many years. What linked these varied pursuits? John Summerson noted long ago the important role that visualization played in Wren's work. Aside from their practical utility, his inventions and discoveries are linked by a powerful ability to envision complex mechanical and geometrical problems in three dimensions and to make those problems intelligible through drawing and model-making. This talent was emphatically not – *pace* Inigo Jones – an attempt to mimic the ideal of Italian *disegno*, but was instead an amalgam of technical drawing practices rooted predominantly in the mathematical sciences and their related craft traditions.[16]

This section of our study explores the theme of visualization as it appears in Wren's scientific work in the mid-1660s: that is, during his slow but determined turn toward architecture. The material comes predominantly from the Wren collection at All Souls College, Oxford, in a series of drawings that illustrate his wide-ranging interests and the almost universal embrace of his draughtsmanship. The first is a sheet of exploratory sketches for Wren's self-regulating weather clock (figure 73, catalogue 38). This device was related to his study of physiology and medicine – how, in particular, the variations and properties of the air affected health. What Wren envisaged – under the general heading of the 'History of Seasons' – was to compile data for two distinct sets of natural phenomena. The first was a meteorological record of the qualities of the air, including wind direction and speed, temperature, and pressure, which the instrument would inscribe automatically during a twelve- or twenty-four-hour period. The second was a correlative register of crops and fruit harvests, stocks of cattle, fish, fowl, and especially epidemics and mortality rates. Such a comparative programme, Wren ventured in 1657, would yield 'a true Astrology to be found by the enquiring Philosopher, which would be of admiral Use to Physick'.[17]

We know of at least three versions of this device, comprising several distinct elements, which also went through various permutations. The first arrangement appears in an undated illustration in the 'Heirloom' copy of the *Parentalia*, the second was seen by the French traveller Balthasar de Monconys in Wren's rooms at All Souls in June 1663, and the third was a more compact example presented to the Royal Society on 9 December of that year (figure 74). The last version comprised the three most basic components of the device. On one side was a weathervane perched atop a tall rod (the arrow itself is not shown), which turned a disc inscribed with concentric circles to mark the hours and with quarters and degrees to indicate direction. On the left was a combined thermometer–barometer in the form of a horizontally pivoting drum containing a curved, open-ended brass tube filled with mercury. Expanding and contracting with changes in the atmosphere, the reservoir of air contained in the drum would displace the mercury's centre of gravity, tilting the barrel one way or another. The core of the machine was a large pendulum clock, which pulled a toothed strip attached by a frame to two

74. Christopher Wren, fold-out drawing of a weather clock, from Royal Society Register Book Original, II, 321–22, dated 4 December 1663

moving pencils. These recorded the motion of both meteorological instruments directly onto their respective surfaces.[18]

The All Souls sheet shows Wren exploring possible configurations for at least two components of the device. In the centre is a sketch of an alternative form of weathervane, adapted to measure not only the direction but also the force of the wind. The recording disc is shown fitted with a system of geared wheels, intended to transfer the motion of the vane to a weight suspended by a pulley. To the left of that drawing is a sketch of a similar geared mechanism, this time attached to a horizontal drum, presumably containing Wren's thermometer–barometer. The mechanism also incorporated a toothed wheel and a rotating arrow. These apparatus are shown in elevated, quasi-perspective views, a pictorial convention familiar from the Italian and German *Theatres of Machines* of the late sixteenth and early seventeenth century. Wren would have known the genre, if only through a famous local example: John Wilkins's *Mathematicall Magick*, written by his Wadham mentor.[19] The sheet contains another significant drawing. Within the space of a few months, Wren reused the same paper to record some notes on an early architectural project. On the corner opposite the weather clock drawings is a scaled, annotated plan for half of the chapel screen at All Souls, a project from 1664. The jottings show him calculating the surface area of both the outer screen and the twinned columns of the inner portal, probably for a quick estimation of materials. This striking juxtaposition conveys some sense of Wren's ability to juggle his competing interests.[20]

Another drawing from All Souls uses a very different set of pictorial conventions. The geometrical diagrams in figure 75 (catalogue 39) were prompted by the appearance of a comet in late 1664. They illustrate three different ways of plotting the object's path, following a theory

Compass and Rule

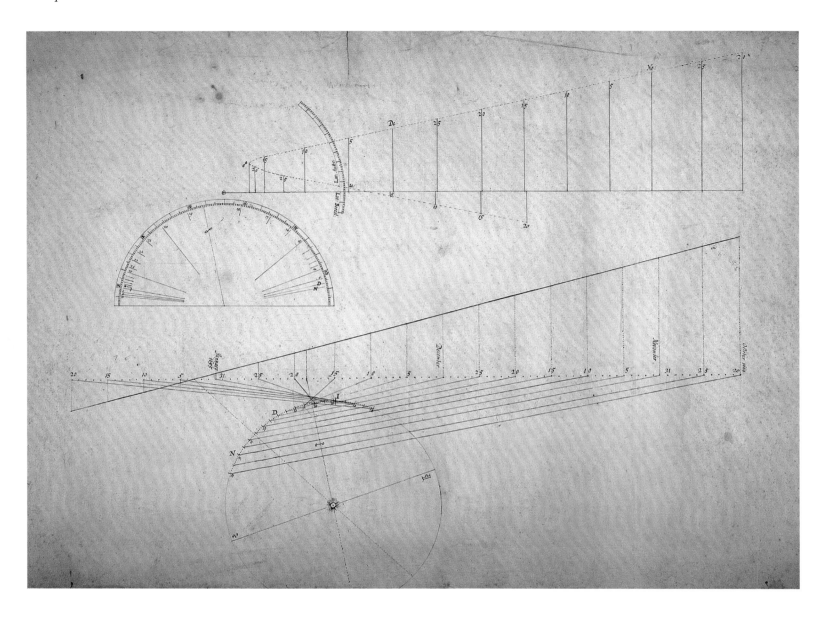

75. Christopher Wren, diagram of the 1664 comet, January 1665. All Souls College, Oxford. Catalogue 39

originally put forward by Kepler that comets moved in a straight line and at a uniform speed. This hypothesis assumed, the comet's trajectory could in principle be determined from just four observations. The sheet of drawings represents the next stage of the problem: how to translate that straight line in space into parameters that could be checked against other observations.

The fundamental diagram is the one at the bottom of the sheet, from which the other two are derived. The circle is a plan view from below of the earth's orbit around the sun, with the earth's changing position plotted clockwise from 20 October 1664 to 20 January 1665. The comet's path is the strong diagonal line running across the sheet, represented in elevation rather than plan. Between the two components of the diagram, Wren has drawn lines at five-day intervals correlating the earth's position with that of the comet. Wren's problem here is to represent the three-dimensional movement of two objects over a given period in two dimensions. If the surface of the paper is understood to represent the ecliptic – the plane of the earth's orbit – the line of the comet begins above the sheet to the right of the diagram at the height indicated by

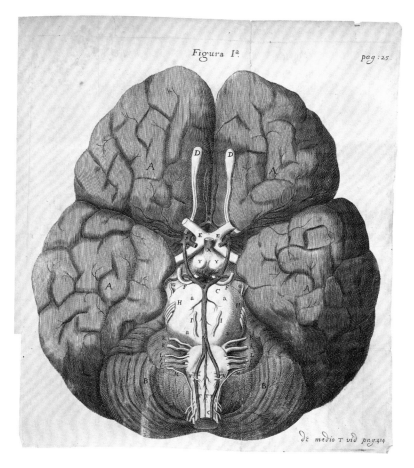
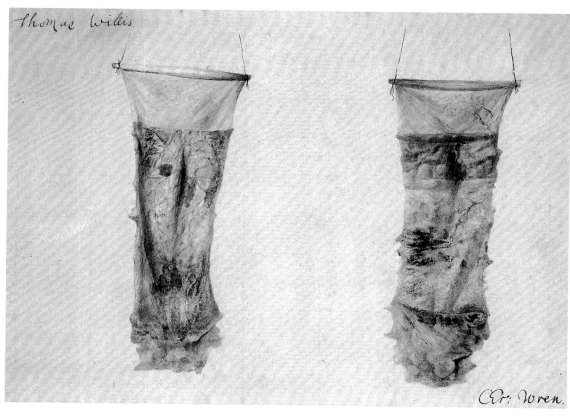

76. Christopher Wren, engraving of the brain, from Willis, *Cerebri Anatome* (1664)

77. Christopher Wren (attr.), watercolour sketch of a piece of small intestine, c.1663(?). Wellcome Library, London. Catalogue 40

Compass and Rule

78. Christopher Wren, clasped hands, date unknown. All Souls College, Oxford. Catalogue 41

the perpendicular. It is then imagined to descend along the inclined path as the earth swings toward it. The two bodies pass each other at conjunction on 18 December, and the comet then continues, intersecting the ecliptic – the sheet of paper – and passing through to the other side on 3 January.

The lower diagram provides a geometrical model of the comet's movement – a 'God's-eye' view, so to speak, of the phenomenon. The two upper diagrams serve to register its observed position as seen from earth. The graduated semicircle records the comet's changing longitude on the circle of the ecliptic. It has been constructed in a purely graphical operation of transferring parallel lines from the lower diagram to the semicircle above. The lines drawn to the perimeter highlight the comet's very rapid shift across the night sky during December, covering almost 150 degrees in a few weeks. The upper diagram on the right was similarly constructed, using compasses to transfer dimensions from below. It acts as a graphical calculator for latitude – that is, the comet's angular distance from the ecliptic plane, represented by the horizontal line with the earth as the little circle at its left extremity. By the simple expedient of using a ruler to join the earth with the comet on its descending line, the comet's latitude is read off where the ruler cuts the circular arc of degrees. The diagram is not a view, but a sort of graph or instrument. The comet's line of descent apparently doubles back on itself simply because it approaches and then retreats from the earth.[21]

Vision, Modelling, Drawing: Christopher Wren's Early Career

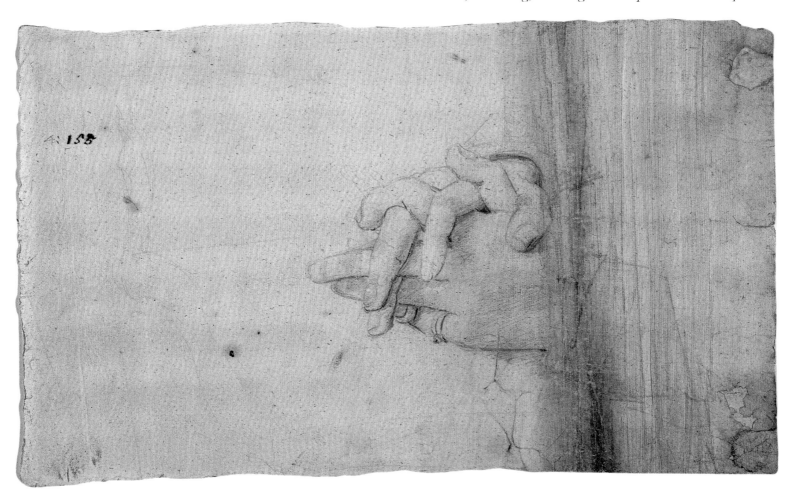

79. Verso of figure 78

This reconstruction takes place in an abstract geometrical space, but this was not Wren's only mode of visualization. He was also adept at direct observation and drawing from life, a skill that he appears to have learned as part of his study of anatomy. The most well-known examples of this talent are his engravings of the skull and brain in Thomas Willis's *Cerebri Anatome* of 1664 (figure 76). It is unfortunate, however, that we have no preparatory drawings for these engravings, which stand at a certain remove from Wren's direct observation of the organs. It is for this reason that the small coloured sketch of an ulcerated intestine, now held in the Wellcome Institute, is potentially so important (figure 77, catalogue 40). If it is indeed Wren's – as the signature indicates – it would represent the sort of preliminary drawing that he might have composed for his engravings of the brain, in a medium that better reflects his careful visual study and his direct grasp of the object of vision.[22] Two other, more securely attributed drawings provide a different example of Wren's ability in this area (figures 78, 79, catalogue 41). These sketches of clasped, resting hands are similarly unique cases in his surviving *oeuvre*, not only as drawings from life but as partial figure studies. They are not without a certain awkwardness, however. Interlaced fingers obviously gave him some trouble, but this, too, points to the fact that Wren's training in this medium was not that of an artist but of a self-taught observer and recorder. In this respect, the drawings offer a fascinating glimpse of Wren in a rare, unguarded moment, directing his gaze for no immediate 'intellectual' goal but purely for pleasure.[23]

80. Christopher Wren, preliminary scheme of the Sheldonian Theatre, 1663. All Souls College, Oxford. Catalogue 42

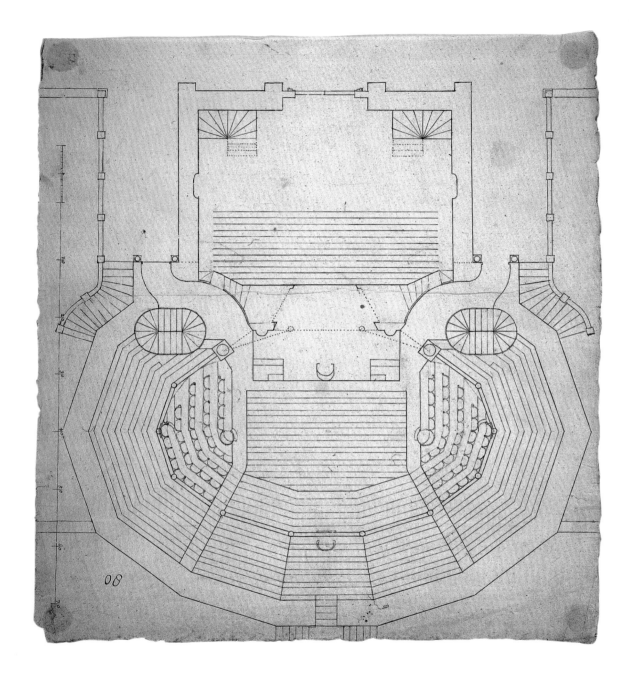

Wren's gift was twofold. First was his 'manual' skill as a draughtsman and modeller, but second – and most important – was his unique ability to apply this talent to ever 'higher' objectives. Put another way, Wren's inventions and scientific insights were not only expressed but were also conceptualized and refined through drawing and model-making. In terms used by Willis, for example, Wren's engravings issued from 'his own most skilful hands', but their usefulness derived from his experience as an anatomist and his understanding of the nature of the organ.[24] Almost all of Wren's scientific projects reflect this combination of skilful means and inventive purpose. Robert Hooke put his finger on this point in a famous and unequalled description of his friend. Wren's microscopic drawings had inspired Hooke's own *Micrographia*, and, in the book's preface, the author acknowledged the debt: 'the hazard of coming after Dr. Wren did affright me;

81. Christopher Wren, pre-fire design for St Paul's, west elevation of the transepts and dome, 1666. All Souls College, Oxford. Catalogue 43

for of him I must affirm, that since the time of Archimedes, there scarce ever met in one man, in so great a perfection, such a Mechanical Hand, and so Philosophical a Mind'.[25]

For Wren in the early 1660s, architecture was a new and fascinating challenge, but it must also have seemed familiar, a pursuit that required an analogous combination of practicality, ingenuity, and inventiveness. Moreover, as Summerson has argued, Wren's early efforts in the field relied on the same sensibility that he had honed through his work as a scientist and mathematician, namely a knack for fusing concept and purpose in visual and physical form. In this respect, the period from 1663 to 1666 represents a turning point in his career, for it was during this time that Wren began to divide his attention between science and architecture. Indeed, his activities in this period were more diverse and intermingled than they would ever be again. Wren's primary medium in the art was – naturally enough – drawing. In the examples we have from these years, we see a talented but uninitiated amateur attempting to master a complex formal language as well as a new set of representational conventions.

We have included three drawings from Wren's early architectural projects, chosen to represent the range of design problems that he set himself. The recently rediscovered plan of the Sheldonian Theatre – an early scheme of 1663 – shows Wren grappling with a complex brief for a major public building (figure 80, catalogue 42). Only his second commission, the Sheldonian was to be the first classical building in Oxford and an explicit attempt to emulate Roman grandeur. A 'Theatre for the Act', it was designed and constructed for the purpose it still serves, as a purpose-built venue for the university's degree-granting ceremony. Like the executed version, the design includes a polygonal auditorium with encircling galleries, and a flat façade with a ceremonial entrance oriented toward the Divinity School. The drawing differs from the realized work, however, in an important respect, namely the inclusion of a raised stage placed between the entrance hall and the auditorium. This feature no doubt reflects the requirements for public examination that formed the core of the ceremony, in which degree candidates disputed formally with a senior member of the university, but its presence also points to a forgotten aspect of the building's original programme: the theatre was intended to house both plays and public dissections. The idea was abandoned soon after Gilbert Sheldon took over the project's financing, reducing the budget and necessitating an overall revision of the design.[26]

For a newcomer to the field, Wren was surprisingly ambitious. In March 1666, Dean William Sancroft asked him to make a recommendation to repair the crumbling medieval fabric of old St Paul's, addressing, above all, the hazardous state of the central tower. Wren was not a member of the official commission in charge of the work, but he nevertheless seized the opportunity to put forward an audacious proposal. Instead of repairing the tower, he would replace the entire crossing with four massive piers carrying a lofty dome. In his accompanying presentation drawings, we see him once again adapting the medium to his purposes, but in what was for him a novel way. His beautiful dome elevation, embellished with blue and gold wash, was only secondarily a 'technical' drawing (figure 81, catalogue 43). Its principal aim was rhetorical, to impress the commission – especially Sancroft – with the proposal's modernity, structural daring, and classical authority. As historians have pointed out, the project drew on Wren's study of continental treatises and on his recent sojourn in Paris, combining references to Bramante's proposal for St Peter's, made famous by Serlio, with the slender double-shelled domes of Jacques Lemercier and other French architects. Wren's intent is clear from his report to the commission and especially from the covering letter to Sancroft, in which he contrasted the 'trew Latine' of classical architecture to the mouldering Gothic of the cathedral, a learned analogy precisely calculated to flatter the patron's taste and humanist sensibility. In a similar vein, the covering

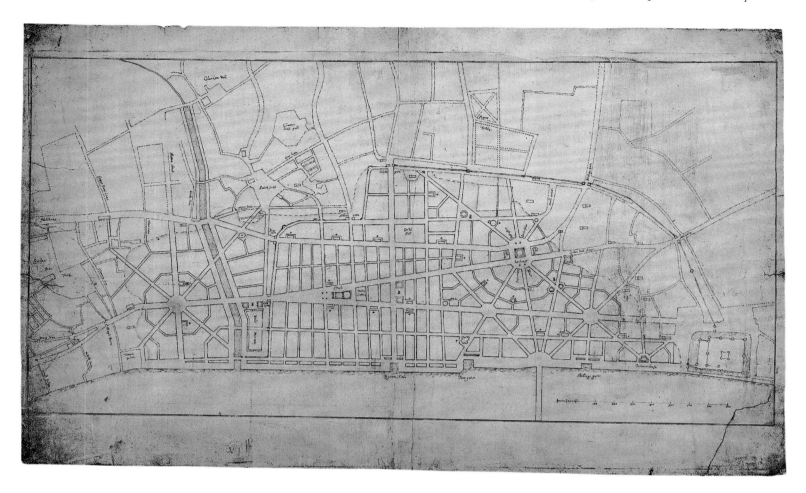

82. Christopher Wren, project for the replanning of London, 1666. All Souls College, Oxford. Catalogue 44

letter of 5 August suggested in what spirit Sancroft was to regard the drawings: 'I have with a great deale of paines finished the designes for it, if they may be usefull, if it happen they bee not thought soe I shall not repent the great satisfaction and pleasure I have taken in the contrivance, which aequalls that of poetry or compositions in Musick'. Wren was careful to present the drawings with a show of unconcern, the product of a gentleman's diversion. 'It hath been', he wrote, deploying what had become a familiar trope, 'my constant Recreation when Journies business or friends left me vacant'.[27]

The drawings were formally approved at an on-site conference on 27 August 1666 and – as is well known – rendered useless a week later. Wren's response to the Great Fire again displayed unanticipated graphic and design ability, this time as an urban planner (figure 82, catalogue 44). Summerson has emphasized the astonishing speed – some five days – in which Wren composed and drew up his plan for the rebuilding of London. In that time, he managed to distil the entire tradition of early modern urbanism and adapt it to the remnants of the destroyed city. A network of broad, straight avenues, like that of Pope Sixtus V, served to link the fixed reference points of the Royal Exchange, St Paul's, the Tower of London, and the north end of London Bridge. Radiating, concentric street patterns drew on ideal city plans and fortified new towns such as Palma Nuova or Villefranche-sur-Meuse. Theatrical set-pieces looked to the Rome of Alexander VII, such as the triangular space opening out from Ludgate to St Paul's – an image of the Piazza del Popolo. The long quays must surely have stemmed from those of Louis XIV's Paris, which Wren so admired. For all its destruction, the fire promised to provide Charles II with a rare opportunity: a reborn capital for a restored monarchy, 'fitter for commerce, apter for government, sweeter for health, more glorious for beauty'.[28]

What we see in comparing Wren's earliest architectural drawings with his contemporary scientific work is not what we might expect. They are not the products of a 'geometrician' or a purely technical expert. Of course, Wren understood *firmitas* to be a basic component of the art, and he was often consulted on that basis. Charles II, for example, had offered him the surveyorship of Tangiers in 1661, and his involvement with St Paul's began the same year. Seth Ward would ask him to report on the state of Salisbury Cathedral in 1668, as would Francis Atterbury at Westminster Abbey in 1713. It is clear, however, that Wren never subscribed to such a narrow definition of the art. His drawings from these early years – even in the limited selection we include here – show him striving to become as universal an architect as Inigo Jones, equally attentive to the rhetorical as to the technical demands of the art. This desire is best seen not as a departure from Wren's scientific identity but rather as an outgrowth of it. This continuity of interest relied partly on architecture's 'mathematical' content, represented, for example, by the use of scale drawing as a planning tool, proportional design methods, and urban cartography. More important was the fact that Wren's notion of 'science' itself was intrinsically visual, concrete, geometrical, and practical.

Chapter Six
Structure and Scale: The Office of Works at St Paul's

How should the relationship between Wren's science and his architecture be defined? Historians have often remarked on Wren's open, responsive approach to design, his willingness, when faced with constricted circumstances or an exacting brief, to depart from established classical and Renaissance precedents. The unusually broad plan of the Sheldonian Theatre or the external screen walls over the aisles at St Paul's are perhaps the most famous examples of this tendency. More striking are the inventive structural features that he sometimes hid within the fabric of a building. The Sheldonian's innovative roof truss or the inverted foundation arches of the Trinity College Library, Cambridge come to mind. Such a sensibility, which might in a loose sense be called experimental, surely owes something to the detachment and empiricism that characterized his approach to scientific questions.

The most spectacular instance of this attitude is also the least understood. Here we explore the final sequence of drawings that determined the shape of the dome of St Paul's and, in particular, that of its canting inner drum. As historians have long suspected, Wren's design for the dome's section was influenced by a novel mechanical theory, namely Robert Hooke's insight that the curve formed by a hanging chain, when inverted, would provide the shape for a 'perfect' masonry arch. In view of this causal relationship, this component of the dome surely represents a landmark in modern structural engineering — one of the first recorded instances in which mathematical science, in a form worked out prior to the design process, was 'applied' to an actual building. That the trial was executed on such a monumental scale is all the more astonishing.

Aside from their purely technical content, the dome drawings also reveal something of Wren's role in the project — the nature of his working relationship with his draughtsmen, for example, or the way he dealt with mounting concerns over expenses. His status closely resembled our own notion of the head of an 'office'. This professional role is familiar because Wren himself helped to create it, but it was largely imposed on him by the pressure of circumstances. The Great Fire of 1666 necessitated a vast array of rebuilding that far outstripped the capacity of the existing building trade. Wren's two design offices at St Paul's and at Whitehall were forced to develop into organizations capable of handling enormous outlays of funds and coordinating hundreds of labourers simultaneously on many different projects, in particular the enormous worksite of St Paul's. The administration would even serve as an alternative training ground for future generations of designers.

One of the innovations that occurred at St Paul's was the institutionalized use of working drawings for individual elements, decorative schemes, and even details. The great wealth of surviving drawings — formerly in the Guildhall Library and now in the London Metropolitan Archives — points not only to the project's size and complexity but also to an unprecedented level of centralization in the design process.

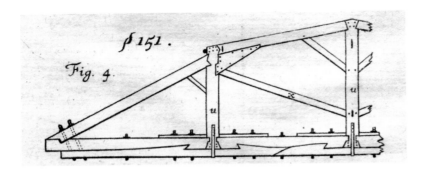

83. Roof truss of the Sheldonian Theatre, from Plot, *The Natural History of Oxford-shire* (1677)

84. Axonometric drawing of Trinity Library, Cambridge, showing inverted foundation arches (drawing by Edward Impey, after an original by Donald Insall Associates)

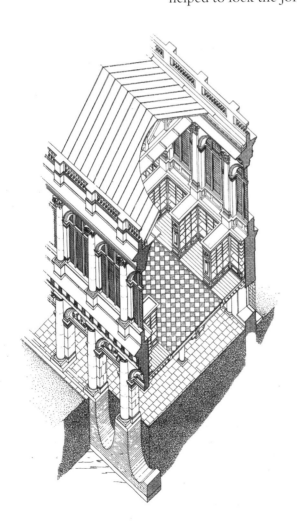

Wren has a special place in the history of structural mechanics. His second 'Tract' on architecture – published posthumously by his son in the *Parentalia* – contains one of the earliest attempts to analyse the behaviour of arches and vaults geometrically. It is important to point out, however, that he did not generally build in this fashion. The use of mathematical calculations as a control over the design process is a modern innovation, one that long postdates Wren's death. Although he often faced structural problems – many resulting from his own ambitious designs – he tended to approach these empirically, turning either to traditional craft means or, more dramatically, to novel expedients and technological solutions. The plan of the Sheldonian Theatre, for example, presented a particularly formidable challenge. At seventy feet, the width of the structure far surpassed the dimensions of a traditional roof truss, the triangular frames repeated along the length of the building to carry the roof. The main problem was finding timbers large enough for the members, particularly the horizontal tie beams. Wren's solution was to create a continuous tie from two layers of seven interlocking beams, scarfed and bolted together so as to transmit the tensile forces (figure 83). Suspended from three vertical posts, the weight of the tie helped to lock the joints of the principal rafters together, as though to form an arch. The system not only served to cover an unprecedented span; it was also sturdy enough to carry Robert Streeter's painted ceiling, suspended below, and to support the university printing presses, which operated from the attic.[1]

Wren's approach to problems of this sort reflected the ingenuity and practical inventiveness of his scientific work – save for one important point. The solutions had to remain hidden, so as not to disrupt or compromise the classical articulation of the building's exterior. The inverted foundation arches, for example, underneath the eastern arcade of Trinity College Library, Cambridge solved a tricky structural problem in just this manner (figure 84). This unusual device served to distribute the weight of the façade evenly, relieving the concentrated load underneath each of the masonry piers, any one of which might be liable to subsidence. Robert Hooke had recently used a similar scheme at Montegue House, but its use here appears to have been extemporized. (The existence of the arches was not suspected until a partial excavation revealed them in 1970.) An analogous artifice, also at Trinity Library, is hidden in the first-floor bookcases. To relieve the excessive weight on the wooden floor, Wren strapped them to the wall with diagonal iron bars. He resorted to a similar expedient at Duke Humphrey's Library, Oxford in 1700, where the weight of the bookcases had opened cracks in the ornate Gothic ceiling of the Divinity School below and, indeed, threatened the whole of the building (figure 85). In a letter to David

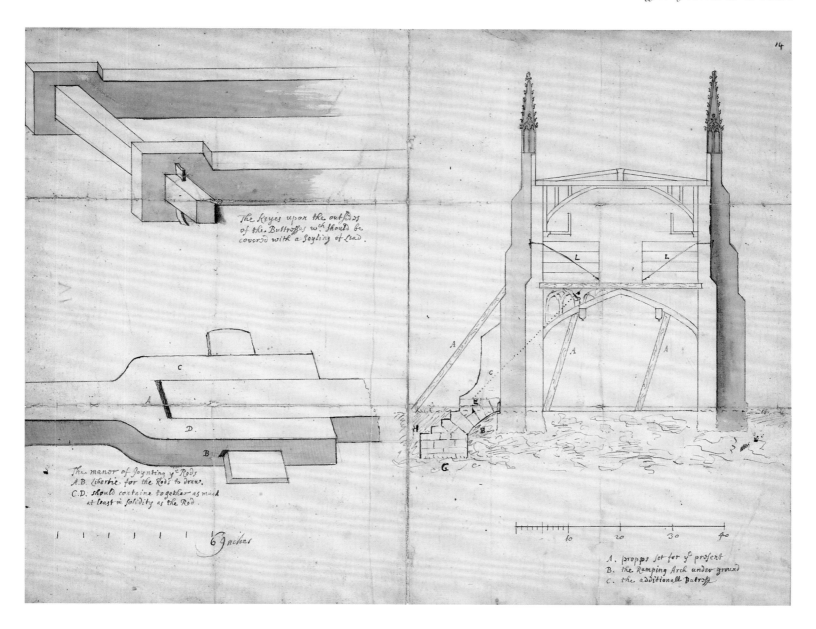

85. Christopher Wren, section of the Duke Humphrey Library, Oxford, with additions and notes concerning repairs and, on left, details for the fabrication of iron ties. Bodleian Library, Oxford MS 907, ff. 13-14

Gregory, then Savilian Professor, Wren recommended substantial use of iron braces to shore up the external buttressing and to tie the bookcases to the wall. His instructions reflect a long and considered experience with the material as a structural support.[2]

This tendency towards improvisation also shines through in the design history of the dome of St Paul's and, to a lesser degree, in the cathedral as a whole. It was apparent in what was previously known of the building's complicated path to realization. The dramatic transitions from the First Model to the Greek Cross to the Great Model to the Warrant Design to the Penultimate Design to the 'Revised' or 'Definitive' design had already told us something of Wren's working method. As a designer, he was responsive to the exigencies of the moment and, in meeting them, capable of rapid and seemingly arbitrary shifts of conception and style. It now appears that the evolution of St Paul's was even more protracted than historians had hitherto believed. Although it has long been known that Wren began modifying the official Warrant

Design both before and after construction began in 1675, the details of this transformation may finally be fitting into place. What they suggest is that Wren started building with few firm or settled ideas for the ultimate form of the cathedral. On the contrary, he continually reassessed and altered the whole design as the building rose, often in response to external factors, such as a change of monarchy, the ebb and flow of income, and the availability of materials.

The abundance of surviving drawings produced during construction poses, paradoxically, one of the chief obstacles to reconstructing its history. The London Metropolitan Archives preserve a bewildering array of proposals, revisions, and variations for different stages of the design, few of which accord wholly with the built fabric. The problem is rendered more acute by the fact that Wren relied so heavily on draughtsmen, which has complicated the task of dating the drawings with questions of attribution. Indeed, the surviving evidence for the development of the dome seemed so intractable that Margaret Whinney flatly declared that it was simply 'not possible to follow the steps by which the final design was evolved about 1704'.[3] Recent research, however, on the handwriting and decorative details of the existing drawings has helped to clarify the situation. In an important article of 2004, Gordon Higgott was able to assign many of these to Wren and to four other draughtsmen, Nicholas Hawksmoor in particular. Fixed in relation to the published building records, the identified drawings may finally provide a reliable chronological framework for the cathedral's progress.[4]

Higgott argued that the favourable circumstances – and increased budget – brought about by the accession of James II in 1685 provided Wren with the impetus to aggrandize the design, giving rise to many of the features embodied in the 'Definitive' or 'Revised' version. Among other changes, the architect decided to enlarge the dome with a tall drum and lantern, modelling it in important respects on that of Jules Hardouin-Mansart's church of Les Invalides.[5] Preparatory work to receive the dome began in 1687, with the masons laying levelling courses of masonry on the tops of the crossing piers, but, by 1690, it must have been apparent that the weight of the structure Wren was proposing would have been too great for the existing supports. In the late 1680s, the facing stone on the crypt piers started to crack as a result of uneven settling and compression on the rubble core. This primary consideration appears to have framed all of Wren's subsequent thought about the dome. The final evolution of the design – evident in proposals datable from 1690–97 and in the built fabric – was driven by the need to lighten the structure and to support the lantern more efficiently and effectively.[6]

The sequence of ideas represented in these drawings reveals Wren's flexibility as a designer and his capacity to respond extemporaneously to developing structural problems. In one respect, however, the dome represents a departure from his normal working method. Unlike his solutions at the Sheldonian or Trinity Library, the dome of St Paul's was not derived in a wholly empirical manner, but rather incorporates the results of recent advances in mechanics. Evidence for this connection comes from a small sketch in the British Museum, datable to $c.$ 1690 and identified – apparently by Hawksmoor himself – as being in Wren's own hand (figure 86, catalogue 45). The drawing shows a proposed triple-shell dome for St Paul's, perched on a tall drum and supporting an ornate lantern. In each of the two matching interior sections, Wren has based the shape of the middle dome on a parabola, determined geometrically as the cube of its distance from the centre. The curve was plotted by aligning the four divisions of the base with the ordinals running down the side of the right-hand drawing in units of eight.

Wren had long been interested in the behaviour of arches and domes. His manuscript 'Tract' on the subject appears to have been part of an investigation dating to late 1670 and early 1671,

when both he and Robert Hooke presented their ideas to the Royal Society on 'the line of an arch for supporting any weight assigned'. The minutes presumably refer to the catenary. A few years later, in 1675, Hooke would publish this discovery as an anagram, usually translated as follows: 'as hangs the flexible line, so but inverted will stand the rigid arch'. In late 1671, Hooke took this idea one step further, transposing his principle from arches to use in domes. At a Royal Society meeting on 7 December, Hooke is recorded as having 'produced the representation of the figure of the arch of a cupola for the sustaining such and such determinate weights, and found it to be a cubico-parabolical conoid' – that is, a 'cubic parabola' revolved around the vertex to form a solid. We do not know why Hooke referred to the curve in this instance as a

86. Christopher Wren, study design for the dome of St Paul's Cathedral, in two half-sections, each incorporating the curve of a cubic parabola, *c.*1690. British Museum, Department of Prints and Drawings. Catalogue 45

87. Nicholas Hawksmoor, half section and elevation of twenty-four-sided dome, St Paul's, c.1690–91. London Metropolitan Archives. Catalogue 46

cubic parabola, which is broader in profile than the catenary and closer to the semicircle. Perhaps he meant it only as an approximation of the latter curve. The mathematical demonstration for the catenary was not published until the early eighteenth century, and it is unlikely that Hooke anticipated it. Alternatively, he may have suspected that the cubic parabola was more suitable for domes, the lower portions of which contain greater outward thrusts – although this hypothesis is complicated by the subsequent remark in the Society's minutes: 'that by this figure might be determined all the difficulties in architecture about arches and butments', which seems to equate the behaviour of the two architectural forms.[7]

Whatever the reason for the change, it is evident that in using this curve as the basis of the drawing, Wren understood it to provide – as far as possible – the form of an ideal masonry dome, one shaped so that it followed, or could at least be shown to contain, the line of thrust. For Wren, this basic principle of efficiency had been a long-standing concern. Although the discussion of arches and vaults in 'Tract II' does not refer to such curves, it nevertheless begins with the question: 'Can an Arch stand without Butment sufficient? If the Butment be more than enough, 'tis an idle Expence of Materials; if too little, it will fall; and so for any Vaulting'.[8] In light of this principle, the most salient feature of the drawing is the way that Wren has sought to give the drum and peristyle the narrowest scope possible to enclose the curve, while keeping them within a more-or-less rectangular section. In subsequent alterations of the design, he would reconsider the relationship between the dome and buttress and, in the process, reduce the drum's section even further.

The British Museum drawing has been known since at least 1923, but it has never been properly studied in the light of the architectural history of the cathedral.[9] Indeed, the evolution of the dome as a whole has remained largely unexplored. In his contribution to this book, Gordon Higgott traces the circumstances that drove the final sequence of drawings for the dome. Wren's thinking regarding this process can now be adequately reconstructed. Perhaps the most surprising result of the series is to see how thoroughly Wren delegated design proposals to his office. The cathedral would not be completed for another twenty-five years, but the British Museum drawing is probably the last we have for any part of it in Wren's own hand. It was not simply a matter of assigning the articulation and decorative details. Wren appears to have given his draughtsmen general instructions regarding the mechanical principles involved and allowed them to explore their own solutions for the structure. This process began almost immediately. After putting the basic idea of the parabola down on paper, Wren handed it to his assistant, the young Hawksmoor, who developed the concept into a series of working proposals. One, inspired by Wren's example, shows him trying to contain the thrust of the dome within the depth of the abutments (figure 87, catalogue 46).

The earliest drawing to show the adopted solution for the whole dome was similarly put forward by a draughtsman (figure 88, catalogue 47). It appears in a composite section, attributable to the office surveyor, William Dickinson, and begun, according to Higgott, in

88. William Dickinson, section of crossing and dome with alternatives, St Paul's, c.1696–c.1702. London Metropolitan Archives. Catalogue 47

1696–1700. The drawing evidently postdates the construction of the sloping inner sides of the dome, which are identical in both left and right variants and which match the built fabric. It must predate, on the other hand, the construction of the peristyle, which does not match the drawing, and that of the three domes, which are shown here in two different combinations. The drawing then underwent a significant revision. Sometime after work had begun on the peristyle in mid-1700, Dickinson altered the left-hand section, introducing, in thick pink lines with diagonal hatches, the basic idea of a straight-sided brick cone rising from the springing of the inner dome, as was ultimately executed.

The drawing is emblematic of Wren's tendency to solve problems as they came to him. For example, the decision to cant the walls of the inner drum must have been made by January 1696, when work began on the base of the drum, but this was well before any decision had been taken for the dome's completion. That is not all. As Higgott observes, the scheme on the right, with a higher dome and lantern, was probably intended as a timber structure, which suggests that Wren was coping with constrained circumstances and trying to delay this final choice until the last minute. Seen in this light, the decision to proceed with the revised variant may indicate that the project finances had been placed, at least temporarily, on a surer footing.

The 'parabolic conoid' did not survive the design process intact, but it does seem to have influenced Wren's decision later in the decade to incline the sides of the inner drum. This shift may appear minor, but it represents a different way of thinking about the structure and the role that the notional line of thrust plays in it. In the executed building, it works not merely as a force to be contained in the dome's section but also as a guide to lightening the structure and helping to channel the outward thrusts downward. It is significant, too, and consistent with Wren's approach to this kind of problem that the slope – about one foot in twelve – is undetectable from below. We do not know how the angle of incline was determined, but it does not appear to have been derived geometrically. W. Godfrey Allen's famous photographs of hanging chains superimposed on the cathedral's section are most likely an interesting conceit, nothing more. Yet the design was nevertheless rooted in a mathematical concept, a novel understanding of the structural behaviour of architectural elements. In putting this theory into practice, Wren inaugurated a new way of using mathematics in the design process, one that would shape the relationship between architecture and engineering to the present day.

The organizational role of the Office of Works at St Paul's was based partly on medieval precedent. Like a masons' lodge, it was located in makeshift rooms hard by the cathedral, built in late 1666 and early 1667 over a demolished range of cloister buildings against the south or west wall of the Convocation House Yard (figure 89). This two-storey building was the administrative centre for the project and the site of all functions essential to its orderly progress, including bookkeeping, tracking and procuring stores, and – not least – drawing. The upper rooms housed the offices for drawing, conferences, and administration, while the lower rooms were used as stores and living quarters. By 1673, the Convocation House itself was pressed into service, partly as a tracing house. Part of the octagonal chamber where the Great Model was kept was partitioned for use as a tracing floor, on which full-scale details of windows or vaulting could be laid out to help shape and assemble stones. The space was also equipped with trestle tables, presumably used to draw the full-scale templates in preparation for the joiners, who subsequently sawed them out and fitted them together.[10]

Such work would have been familiar to any medieval builder, as would many of the functions of the office, but in other ways its role and organization were unprecedented. There is the obvious point that it was headed not by a master mason but by a gentleman amateur. In

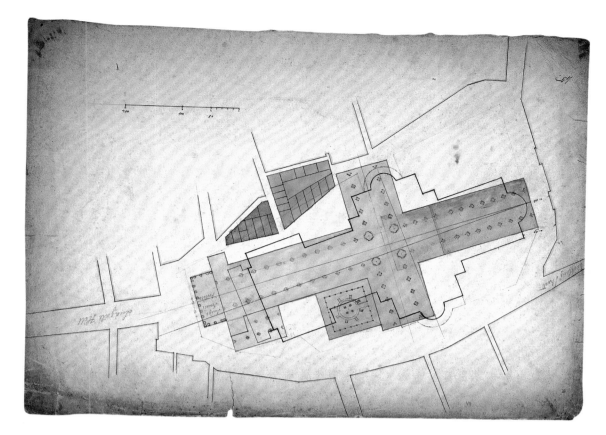

89. Plan of old St Paul's with new cathedral overlaid (showing a slightly incorrect alignment). All Souls College, Oxford, III.45 (Geraghty 109)

fact, few members of the office came from a background in masonry or carpentry. In place of a single craftsman-overseer charged with administering and executing the whole project, the office coordinated the work of multiple mason-contractors, who each brought with them their own crews of men to work simultaneously on different segments of the project. This procedure was implemented partly by necessity – the project was simply too large for a single contractor to handle – but it also allowed greater external supervision over both the design and the building process. Indeed, what stands out most about the way that the office worked was the control – evident in the unprecedented use of drawings – that Wren and his draughtsmen exercised in virtually every aspect of the design. There are more extant drawings for St Paul's – 68 in the collection of All Souls College, Oxford and 220 at the London Metropolitan Archives – than for any earlier architectural project in England. Moreover, the surviving drawings probably represent only a fraction of those originally made. In this respect, Wren was continuing a precedent that he had initially developed for Trinity College Library. Frequently absent from the site, he developed a routine of sending scaled-up working drawings of details and mouldings for the workmen to follow (figure 90). Small features such as architraves and cornices were typically decided by the masons on site, but this was clearly an area of autonomy that Wren sought to wrest from them.[11]

Among other innovations, this state of affairs forced a series of extraordinary advances in technical drawing. These novel representational conventions must have arisen from the need to depict complex physical conditions of the structure and to convey that graphical information quickly and efficiently. Kerry Downes has signalled one such example, which appears on a set of four drawings for the upper transept ends (figure 91, catalogue 48). The left side shows half of the second-storey elevation, coupled with its associated ground plan. On the right side, the

Compass and Rule

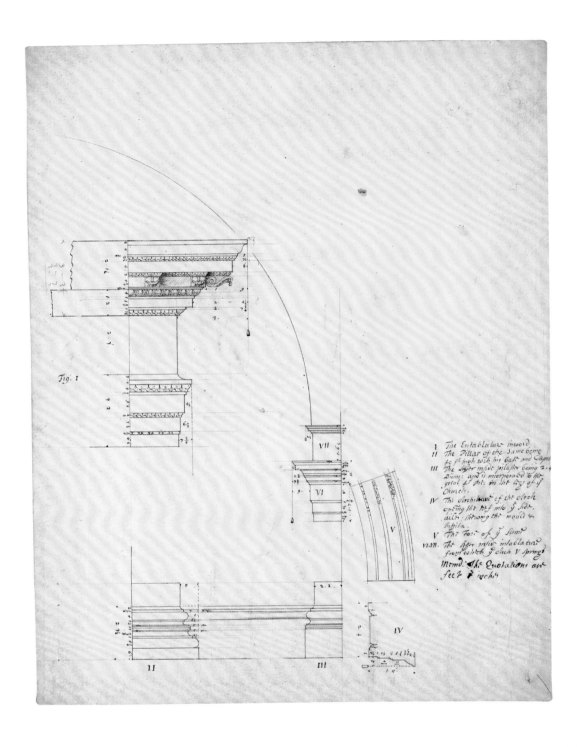

90. Nicholas Hawksmoor (possibly after an original by Christopher Wren), details of internal orders, St Paul's, c.1685–86. London Metropolitan Archives SP 75 (Downes 115)

transept and half-dome of the porch appear in section, along with the internal elevation of the adjacent clerestory window. All four elements are drawn to the same scale and aligned on the page. Finally, below, is an alternative for the internal window surround. While making the drawing difficult to read, the economy of means has allowed the draughtsman to compress considerable information into a very small area. Like a system of visual shorthand, the convention depends on an informed viewer. Where did this technique come from? Downes has proposed that Wren saw examples of it on his travels. There is a suggestive French parallel, for instance, in a contract drawing by François Mansart for the Church of the Visitation, which uses similar

combined views in a compressed format. But Wren did not apparently make the transept drawings. Indeed, Higgott has assigned them to Hawksmoor and to the fourth 'unidentified' draughtsman. Rather than a continental influence, the convention may reflect a common source in the training of masons, who often used combined and overlapping partial views, either on paper or at full scale.[12]

The same two draughtsmen combined to produce another exceptional and even more spectacular drawing: two variant plans for the drum and peristyle superimposed on a quarter plan of the crossing (figure 92, catalogue 49). As Downes has remarked, the drawing was

91. 'Unidentified' draughtsman, with Nicholas Hawksmoor, study for upper transept end, St Paul's, c.1685–86. London Metropolitan Archives. Catalogue 48

Compass and Rule

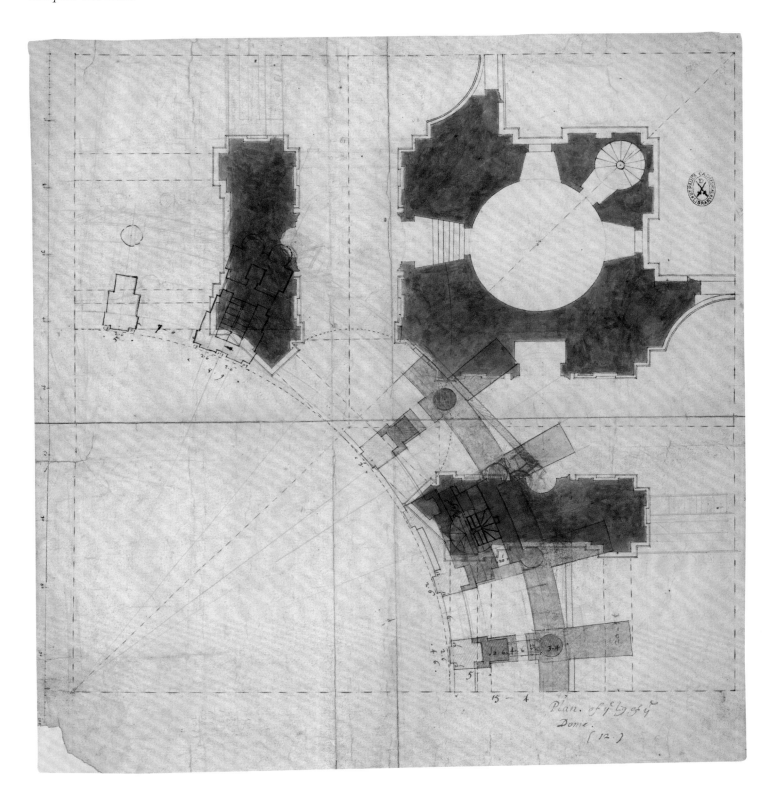

92. 'Unidentified' draughtsman, with Nicholas Hawksmoor, quarter plan of crossing with variants for peristyle, St Paul's, c.1675–87 (crossing plan) and c.1690–94 (variants for peristyle, by Hawksmoor). London Metropolitan Archives. Catalogue 49

probably made in two stages. The plan of the crossing belongs to a related group from an earlier period, which Higgott has dated roughly to 1675–87 and assigned to the 'unidentified' draughtsman. The dashed lines bridging the nave piers represent the 'soffit-bands' of the crossing arches, the chief structural features that conditioned the subsequent design of the drum and dome. The second layer was added later, probably by Hawksmoor, at a fairly advanced stage in the development of the peristyle but before its form and dimensions had been finalized.[13] Of primary interest is the way in which the drawing attempts to represent simultaneously rising vertical stages, not only between the drum and the crossing below but also between different stages of the drum itself. A comparison with the dome's section (figure 138) makes clear that the areas washed in blue represent the lowest part of the drum, namely the sloped radial buttresses that sit directly on the crossing arches and end below the roofline. From a point above this level rise the interior pilasters, which are shaded with cross-hatching, while above these stand the feet of the exterior peristyle columns, washed in yellow. Hawksmoor was trying here to stretch the conventions of the ground plan to ends that they were not really able to serve. For all the drawing's visual interest, it could convey little information about such a complicated physical structure. That is probably why the office relied instead on large-scale masonry models during this stage of the design. The accounts for 1690 to 1695 mention a number of models for the dome, and Hawksmoor's plan may have been prepared for one of them.[14]

The problem remains of how these models and drawings were used to direct the workmen charged with executing the design. As Kerry Downes has speculated, many of the drawings made for St Paul's were no doubt destroyed during the construction process. Since few of the existing drawings match the built fabric, we might conclude that construction drawings predominate among the missing. One survivor, however, suggests what such construction drawings might have looked like (figure 93, catalogue 50). This large-scale section of the peristyle is almost five feet long and bears several inscribed dimensions. The combination of its great size and tight focus on a single structural element is very unusual among the surviving corpus and points to the possibility that it was made as a guide to help lay out a larger version on the tracing floor. As Gordon Higgott observes, however, the drawing differs in an important respect from the built fabric. The fact that the upper tier of windows is left open both signifies that it predates the final decision to introduce the conical brick dome and underlines – if this was indeed a construction drawing for the fabric – the urgency of that decision.

It is not easy to link Wren's science with his career as an architect, even if we accept that the two fields were then understood to share a common paradigm. The fact is that he never found it necessary to connect them explicitly, to articulate the way in which his varied pursuits were meant to cohere. That has

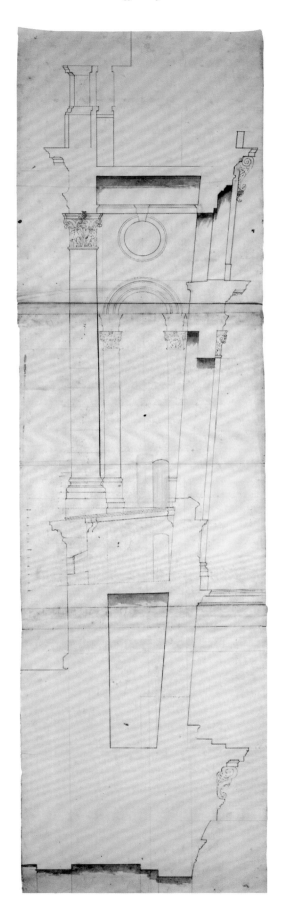

93. Nicholas Hawksmoor, large-scale section through drum and peristyle, c.1700. London Metropolitan Archives. Catalogue 50

not, however, kept historians from being tantalized by the subject. Why would a scholar and philosopher of European-wide reputation trade one career for another in architecture? The question has been the most compelling of his biography, ever since John Summerson's pioneering 1936 article, 'The Mind of Wren', and it continues to drive modern treatments of his life. The value of the British Museum drawing lies in the fact that it is one of the only direct pieces of evidence we have that links his science with his architecture in a practical context. In this regard, it is characteristic of his approach in both fields. As a piece of design, it is improvised and daring, and, as an application of science, it is practical and ingenious. The same combination of qualities recurs throughout his work.

Chapter Seven
Gentlemen, Practitioners, and Instrumental Architecture

The 1666 Fire created the opportunity for Christopher Wren's most important commission. It also led to a huge wave of smaller-scale rebuilding and a corresponding increase in architectural publication.[1] Some new texts remained firmly within the tradition of quantity surveying represented by Leonard Digges's *Tectonicon* of a century earlier. The London surveyor William Leybourn, for example, issued just such a work on mensuration only one year after the fire. But Leybourn also recognized a new convergence of interests between two typically separate classes of builders: 'this Treatise may be beneficial and usefull as well to Gentlemen and others (who at this time may have more than ordinary occasion to make use thereof, in the Rebuilding of the Renowned City of London,) as to Artificers themselves, for whose sakes chiefly it was intended'.[2]

Publishers sought to engage this wider and increasingly diverse audience of gentlemen and mechanicians by combining commercially proven practical mathematics with the less familiar architecture of the orders. The results of this approach sometimes resembled a shotgun marriage more than a seamless integration. A characteristically disjointed but successful example was *The Mirrour of Architecture*, a text first assembled by the publisher William Fisher in 1669, which went through ten editions before 1752 (figure 94).[3] The five orders presented in the first section were based on book six of Vincenzo Scamozzi's *Idea della architettura universale* (1615), but their route to English was a circuitous one: the source was a Dutch abridgement of the Dutch folio translation. Scamozzi was thus reduced from his role as erudite expositor of the art to little more than a pattern-book source.

A completely distinct second section provided a manual by the London instrument maker John Brown on a calculating device specially adapted for structural carpentry. Brown's 'joint rule' was a simplified form of sector, an instrument that had rapidly risen to prominence since the beginning of the seventeenth century and would remain fundamental to practical mathematics into the later nineteenth century. Brown presented his

94. Title page, from Scamozzi, *The Mirrour of Architecture* (1669)

Compass and Rule

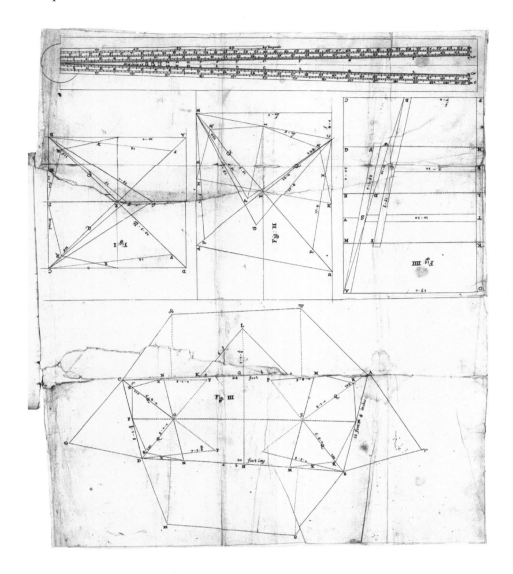

95. Folding plate with joint rule and roofing diagrams, from John Brown's 'The Description and Use of an Ordinary Joint-Rule', in Scamozzi, *The Mirrour of Architecture* (1669)

96. Detail of a sector in use from the engraved title page of Gunter, *The Description and Use of the Sector* (1636)

instrument first as a general tool for proportional calculation and then as a device for calculating the dimensions and angles of the timber-work of roofs. Plunging directly into the detail of hips, rafters, king-posts, pitches, and bevels, he assumed a high degree of craft knowledge: both his text and folding plate would have been extremely challenging for those outside the carpentry trade (figure 95). The change of focus in the last problem is, therefore, all the more surprising: 'The Use of the Scales to lay down or measure out on Paper, or Board, the Members and Parts of the five Columns, and their Ornaments, with their names and measures, digested into a Table, for the more ease and use of Workmen'.[4] In a brief summary, Brown provided a table of the names, heights, and projections of the parts of the Tuscan order, before noting that the other four orders could be easily and similarly supplied.

John Brown's short concluding instructions opened a century-long effort to fix the five orders in instrumental form. The sector was the basis of the majority of these attempts, so it merits a word of explanation before considering the texts and instruments of later practitioners. The instrument is a hinged rule that operates on the principle of similar triangles. It is invariably used with compasses or dividers to transfer dimensions to and from the matching pairs of scales on its two legs (figure 96). The technique is simple but powerful, making the sector a general-purpose tool for calculation (figure 97). By the addition of specially adapted scales, the instrument could also deliver useful results in many branches of mathematics, from navigation to geometry. The typical form of the instrument in England was the Gunter sector (figure 98). John Brown's version stripped away most of the scales to focus on its specific building purpose,

Gentlemen, Practitioners, and Instrumental Architecture

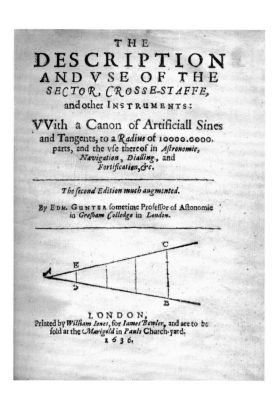

97. A simple example of the use of the sector, represented schematically as AB and AC in this diagram from the title page of Gunter, *The Description and Use of the Sector* (1636). To multiply a length by four, set the dividers to the given length DE. Place one point of the dividers on division '1' on leg AB (using the sector's simplest scale of equal parts, or line of lines). Then open the sector's other leg AC until the other divider point falls on the equivalent '1' mark. Hold the sector legs at this angle and open the dividers until their points fit on the '4' divisions on each leg. The dividers now measure a length four times greater than the original: AB is 4AD, so BC is 4DE.

98. Engraving of the Gunter sector by Elias Allen, used as an advertisement for Walter Hayes and Anthony Thompson, from *The Workes of Edmund Gunter* (1662)

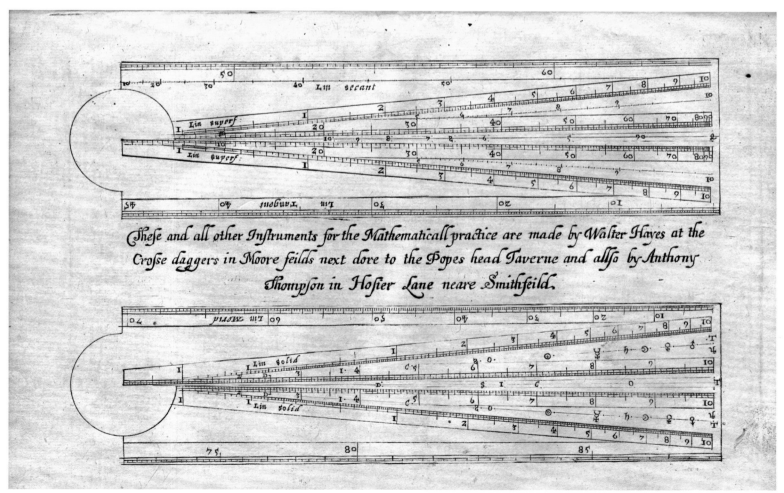

while not losing its powers of general calculation. He also abandoned the name, doubtless because 'joint rule' would sound reassuringly familiar to workmen.[5]

Fisher's 1669 compilation represented an attempt to create a broad, if not entirely cohesive, vision of architecture, by adapting and combining continental and English sources. John Brown's inclusion in the volume suggests that the intellectually dynamic community of mathematical instrument makers played a role in the assimilation and public dissemination of classical architecture as an English art. The nascent trade of the sixteenth century, with makers such as Thomas Gemini and Humfrey Cole, had been tiny. During the eighteenth century, London emerged as the most important world centre for a trade that, through navigation and surveying, was considered vital to commerce and territorial possession. The state significance of mathematics had been signalled by the foundation of the Royal Mathematical School (1673) and the Royal Observatory at Greenwich (1675). London's Restoration wits might laugh at the Royal Society's natural philosophers for studying lice and fleas or weighing air. As Thomas Shadwell satirically jibed in *The Virtuoso*, what was the moral and practical value of such activity? By contrast, mathematics was portrayed as both pleasurable and effective, and instruments provided proof (or reassurance) that it dealt with solid and substantial matters – that it was not merely theoretical speculation but geared towards action.

The mathematical makers had social as well as technical and commercial skills. Their workshops were sites of 'intelligence', where contacts and conversation were traded as well as books and instruments.[6] Brown himself often appeared on his title pages not as a mere tradesman but as a more ingenious 'Philomath' and, although he addressed workmen, his customers also included such men as Samuel Pepys.[7] For gentlemen and artisans alike, the acquisition of mathematical skills through instruments was understood not as a tedious chore but as a source of delight.[8] Brown's successors would continue to incorporate architecture into their entrepreneurial efforts. Whether considered useful for trade or as a polite accomplishment, the latest innovations could be sought out in the fashionable shops of the eighteenth-century West End. This chapter explores the process by which architecture and mathematics were joined through instruments.

Roger North is a particularly striking example of how such instruments might appeal to a gentleman with an interest in architecture. North was the youngest son of the fourth Baron North and followed his older brother Francis into a career in law and politics. As mentioned in Chapter 5, after a fire in 1679, he helped to rebuild the Middle Temple and even redesigned his own chambers. He subsequently oversaw the construction of the Temple Gate in 1683–84. He was also a frequent visitor, with his brother Dudley, to the building site at St Paul's, where he often found Wren.[9] After political displacement following the accession of William and Mary in 1689, he acquired the country estate of Rougham in Norfolk, remodelled the house there, and retired from public life. From this point, writing became his primary occupation. North composed and revised thousands of manuscript pages on subjects ranging from music and mechanics to optics and natural philosophy, along with biographies of three of his brothers and his own autobiography. Architecture, too, had a place in this energetic literary production. Although never published during his lifetime, North's architectural treatise nonetheless stands as a significant contribution to the English literature inaugurated by Henry Wotton's *Elements of Architecture* (1624).

North's autobiographical notes provide a remarkably intimate account of his own intellectual development and the way in which he framed his architectural interests in relation

99. Henry Wynne, magazine case of mathematical instruments, late seventeenth century. Jesus College, Cambridge. Catalogue 51

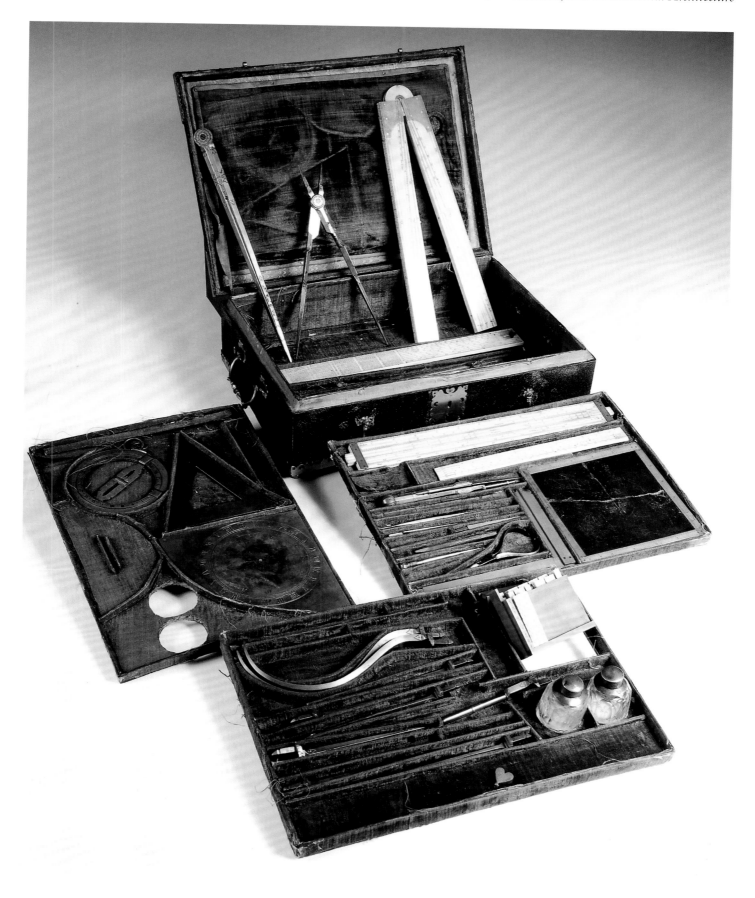

Compass and Rule

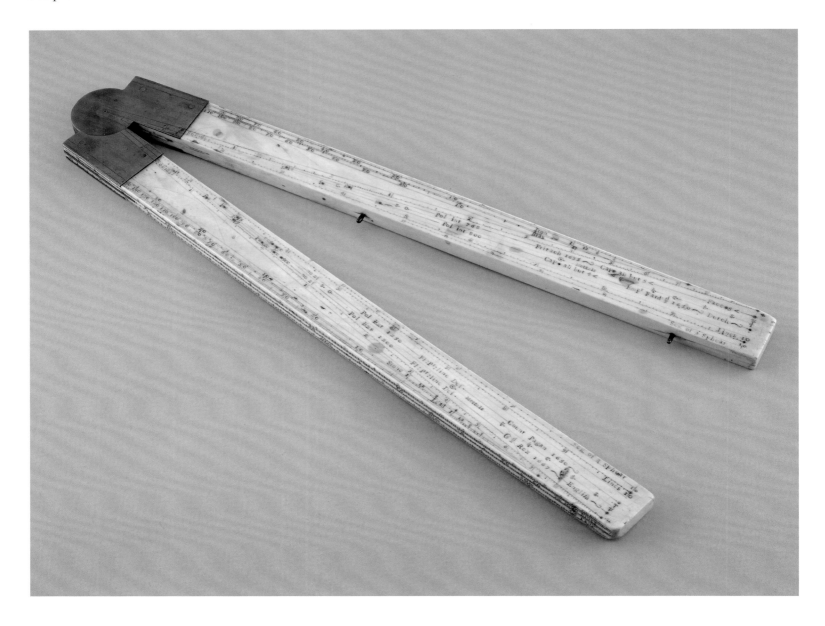

100. Fortification sector to the design of Jonas Moore, *c.*1673. British Museum. Catalogue 52

to mathematics. Even before going to Cambridge as a student, North had been introduced to practical mathematics within the family. A youthful interest in sundials was encouraged by his brother Francis, who gave him a ruler, a pair of compasses, and a text on practical geometry. In his year at Cambridge, North was greatly taken with Descartes, who was then both fashionable and controversial within the university. He also delved into Euclid, but his preference was for practice over demonstration. Subsequent 'practicall diversions' in mathematics included not just perspective and mechanics but even sailing. North linked the 'joys I had in mathematicall exercises' with his other great passion, music. He mused that 'the very remembrance of these things is delight, and while I write, methincks I play. All other imployments that filld my time goe on account of work and business; these were all pleasure'.[10]

The pleasures of mathematics may seem alien to modern readers, but they were real to North. It was not the austere beauty of rigorous proof that attracted him but ingenious accomplishment and practical action. As with music, the most direct experience of that pleasure came in instrumental performance, in the deft mastery of difficult tasks. North keenly felt the responsibility of graver civic and family duties, but he persuaded himself not to feel guilt at their occasional neglect. His recollection of the mathematical recreation of drawing is telling: 'with this I did entertein my self many hours, which might have bin better and more profitably imployed, but I had not power to resist my genius, and flattered my self that not running into vice I was absolved'. Drawing was central to North's architectural endeavours. He did not have the early experience or instruction to render him fit for figurative work, but 'for the regular part of the designe, so as to give the true profile, in proportion all manner of ways, none was readyer then my self, nor more exact'.[11] The pride in exactitude signals the mathematical virtues that underscored architectural drawing and that were embodied in instruments. North's autobiography includes commentary on the merits and uses of specific drawing instruments, where his evident working experience and easy familiarity shine through. He bequeathed two cases of instruments and one of them, an extraordinarily elaborate and complete magazine case, still survives in Jesus College, Cambridge (figure 99, catalogue 51).[12]

North not only bought and used instruments but also designed them. Having begun architectural drawing and acquired some books, he then 'fell into a humour of contriving new instruments'. He set out, in particular, to capture the five orders:

> I also caused to be made an archetonicall sector, whereby one might draw any order after Paladio, to the minutest parts without help of any book or memoriall but what was there … One may with the help of this sector, a square, and a compas sett downe, describe, or make a patterne for workmen, without recours to a book, which no man otherwise can undertake for.[13]

No example survives of this 'architectonic sector' – as such instruments would later be known – but North's description makes it clear enough what it was. One side was a simplified version of the commonly available sector. Rather than having a variety of scales for various geometric tasks, there seems to have been only a single scale of equal parts, specially divided to make it more appropriate for modules ('models') and minutes. The reverse had a table of parts of the five orders from which dimensions could be taken, so that the sector consisted of a geometric means of rescaling in proportion on one side and an aide-mémoire on the other.

North's sector is very strongly reminiscent of John Brown's joint rule of 1669. Both have a simplified layout that contrasts markedly with other specialist sectors available at the time. Jonas Moore, for example, had published a sector adapted for military architecture in his *Modern Fortification* (1673) but with an entirely different and far more elaborate set of scales (figure 100, catalogue 52). Moreover, North's provision of a Palladian table on his sector appears as a direct response to the final problem on the orders in Brown's text, where the table remained in the book but the instrument's reverse was blank. North even mentions in his autobiographical notes that he purchased 'Sciamozzi' among other architectural books, which probably refers to William Fisher's readily accessible *Mirrour of Architecture* rather than the weighty Italian original.[14] But, even if not directly inspired by Brown's contribution to the *Mirrour*, North was a customer of the London instrument shops and had direct contact with the world that Brown inhabited: 'squares, rulers, sectors, and other such convenient utensils, are ordinary and found in shopps, to which the artist, as I was, will be led by occasion, and may fitt himself there according to his fancy'.[15]

Compass and Rule

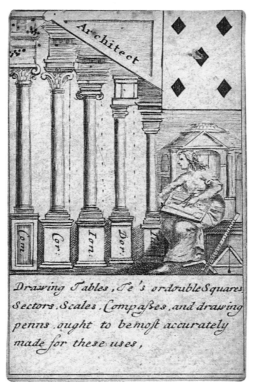

101 a–e. Thomas Tuttell, mathematical playing cards, 1701. British Museum. Catalogue 53

Gentlemen, Practitioners, and Instrumental Architecture

102. Samuel Saunders, portable case of drawing instruments, 1730s. Collection of Howard Dawes. Catalogue 54

If North offers an exceptionally well-documented individual case of the integration of architecture with mathematics, the conjunction is prominent in textbooks too. For example, Joseph Moxon's mathematical dictionary of 1679 blandly asserted that 'Architecture, is a Mathematicall Science, which teaches the Art of Building, or a skill obtain'd by precepts of Geometry, &c.'.[16] But the most amusingly vivid evidence comes from Thomas Tuttell, mathematical instrument maker to the king, who issued a set of mathematical playing cards in 1701 (figures 101 a–e, catalogue 53). His advertisement emphasizes the instrumental character of mathematics and the broad audience he hoped for:

> Tuttell's Mathematical Cards containing all the Instruments generally us'd in Navigation, Surveying, Dialling, Gauging, Fortification, &c. useful to our curious Nobility and Gentry; also Engeneers, Mariners, Gaugers, Gardeners, Builders, Shipwrights, Bricklares, Stonecutters and all Artists (as easie to play with as any) Price 12 pence the Pack.[17]

119

Compass and Rule

103. Thomas Heath, architectonic sector to the design of Thomas Carwitham, c.1723. Museum of the History of Science, Oxford. Catalogue 55

Like the instruments they pictured, the cards aimed to combine pleasure and polite instruction. The ace of hearts introduces the mathematical arts in general, which, it is claimed, 'by the Assistance of a Master are made so delightfull & pleasant & of Such general use, that they aggrandize a Man for all Conversation & make him Capable of any Employ'. Other cards feature individual instruments – the queen of clubs as a drawing table – as well as the personification of trades, with architecture and building featuring strikingly among the arts represented. Whereas the carpenter, glazier, bricklayer, and others are shown as contemporary workmen, the architect is a classical goddess, posed in front of the five orders. This might be taken as a conventional deferral to the classical prestige of architecture, which is also emphasized by the king of clubs as 'Building'. But it could also be read as suggesting the uncertainty and instability of the public identity of the architect. Is it just by accident that Tuttell shows the architect (the five of diamonds) worth exactly half the value of the bricklayer (the ten of diamonds)?

Such sensitivities may indicate a concern to rein in the overweening architect or surveyor, but they did not threaten the routine identification of architecture as mathematics. That trope was even realized in material form. A portable case of drawing instruments from the 1730s by the London maker Samuel Saunders carries a badge of 'ARCHITECTURA' on one side and the twinned titles 'GEOMETRIA PERSPECTIVA' on the other (figure 102, catalogue 54). It was perhaps intended as a gift, and the elaborate gilt decoration of putti and instruments creates an overwhelming statement of conventional wisdom at the luxury end of the market.[18]

The key figure in the further development of the tradition of architectural instruments was one of Saunders's contemporaries. Thomas Heath was a prominent maker, with a reputation for accuracy. Free of the Grocers' Company in 1720, he established himself in the Strand, taking on his former apprentice (and son-in-law) Tycho Wing as a partner in the early 1750s. Working in silver, brass, ivory, or wood, he offered a wide retail range, from the most expensive of grand orreries to more mainstream material such as sundials and instruments for surveying and navigation. Although his shop also stocked books and charts, mathematical instruments were the core of both his business and his reputation.[19]

Between the 1720s and 1740s, Heath appeared as publisher or co-publisher of four books presenting mathematical instruments for architecture. Each volume was by a different author, and it is Heath's appearance in the imprint that links the works together.[20] To be listed as publisher does not necessarily mean that Heath provided the capital necessary for the editions; the primary commercial risk was more likely to lie with the author.[21] But, even if he was not orchestrating and directing the labours of others, it is clear that Heath was not simply a hired mechanical hand. He provided the retail outlet for both books and instruments and probably invested his own time and resources in refining and developing his authors' designs. Heath's consistent engagement is in itself remarkable for an individual maker, and strongly underlines the value of following the maker's perspective. The books and instruments resulting from this effort represent a unique constellation within the wider European context. Nowhere else was there such a concentration in the innovation and manufacture of mathematical instruments for architecture.[22] Although it is tempting to link Heath's enterprise with the Palladian revival of the 1720s, the books and instruments are diverse and do not have a clear programmatic identity. The practical mathematical tradition, whose gentlemanly, mechanical, and commercial values were already established in the seventeenth century, provides a more substantial and secure interpretive context, particularly because Heath had a direct connection to that earlier tradition through a redesigned form of John Brown's joint rule.[23]

Heath first ventured into architectural territory as the publisher of a slim text by the painter and draughtsman Thomas Carwitham.[24] Published in 1723, Carwitham's text presented a new sector design, along with a variant, the 'architectonick sliding plates'. Carwitham's sector differed from the standard type by having three rather than two hinged legs, allowing two sets of proportional operations to be conducted in parallel (figure 103, catalogue 55). Like Brown's and North's before it, his sector could be used to rescale the elements of the five orders in proportion, but the instrument also had a more specific purpose. Its main feature was a set of four scales that permitted the delineation of the flutes and fillets of pilasters and columns in both plan and elevation (figure 104). That this subsequently appeared too narrow is suggested

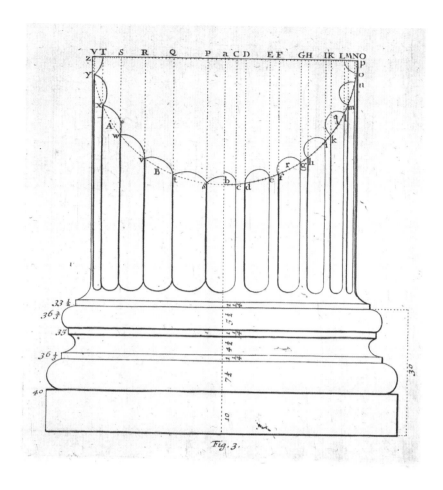

104. Fluted column with and without fillets, from Carwitham, *The Description and Use of an Architectonick Sector* (1723)

Compass and Rule

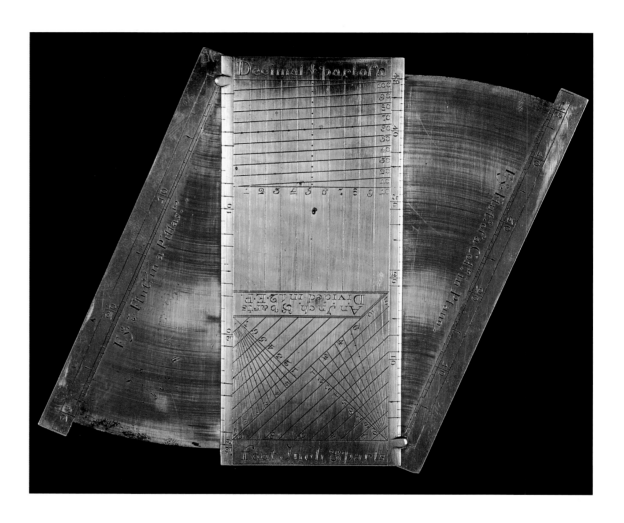

105. Thomas Heath, architectonic sliding plates to the design of Thomas Carwitham, c.1723. Collection of Howard Dawes. Catalogue 56

by the book's second issue in 1733, which sought to broaden the instrument's range by enabling it to draw scale plans and elevations. The architectonic sliding plates presented the same four scales in an alternative format. Used with a special pair of curved compasses, they were intended for use with small dimensions, which would come awkwardly close to the joint of the sector. The device is not illustrated in the text and its exact form would have remained obscure but for the survival of a single example (figure 105, catalogue 56).

Whereas Carwitham never appeared in print again, Heath's next collaborator was the prolific William Halfpenny. His *Magnum in Parvo, or, The Marrow of Architecture* (1728) was one of five works that he authored between 1724 and 1731, all providing simple methods for ordinary builders to master the five orders and other architectural elements (figure 106, catalogue 57). Heath's interest lay in the author's ambitious attempt to instrumentalize and automate the delineation of the five orders with a specially adapted drawing board.

Halfpenny's apparatus was essentially a rescaling device, elaborate and unique in form but simple in principle. Imagine drawing the total height of an order as a vertical line, while marking

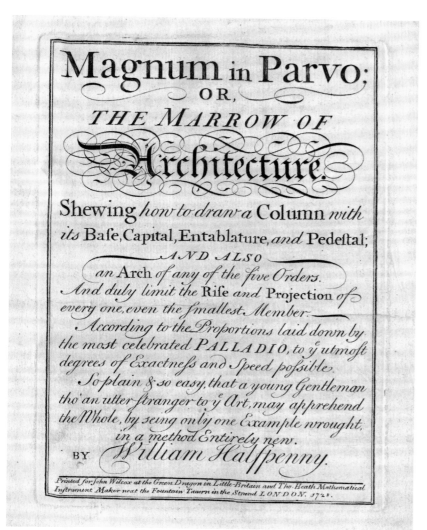
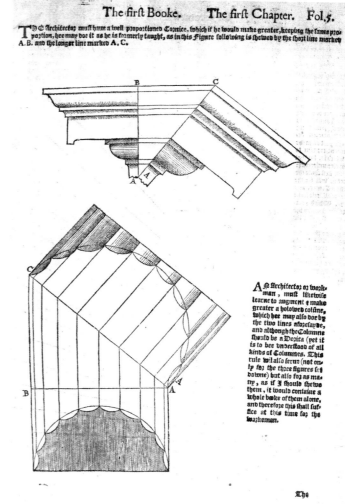

off the height of each element with a series of points along its length. Rotate that line from the base, while drawing a horizontal line through each point. The intersection of those lines with the original vertical provides a smaller version of the order, now reduced in proportion. The amount of rotation determines the extent of the reduction. This technique had long been known: Serlio, for example, illustrated an analogous version (figure 107). Halfpenny's innovation was to turn this principle into material form, with the help of two brass plates placed alongside the drawing surface. The plates carried scales embodying the proportions of each of the five orders, one for the heights and the other for the projections. By rotating these plates, the elements of the orders could be rescaled and transferred onto the drawing with a T-square. The larger device, with a circular segment moving in a semicircle, sits in a groove to the left of the drawing and provides the heights. The smaller plate slides along the bottom of the drawing and has a rotatable double-sided circle within a square frame (figure 108, catalogue 57).

Like Carwitham's instruments, these plates are extremely rare – only a single pair is known to survive, with the large semicircular plate signed by Heath (figure 109, catalogue 58). The

106. Title page, from Halfpenny, *Magnum in Parvo* (1728). Collection of Howard Dawes. Catalogue 57

107. Technique for rescaling, from Serlio, *The First Booke of Architecture* (1611)

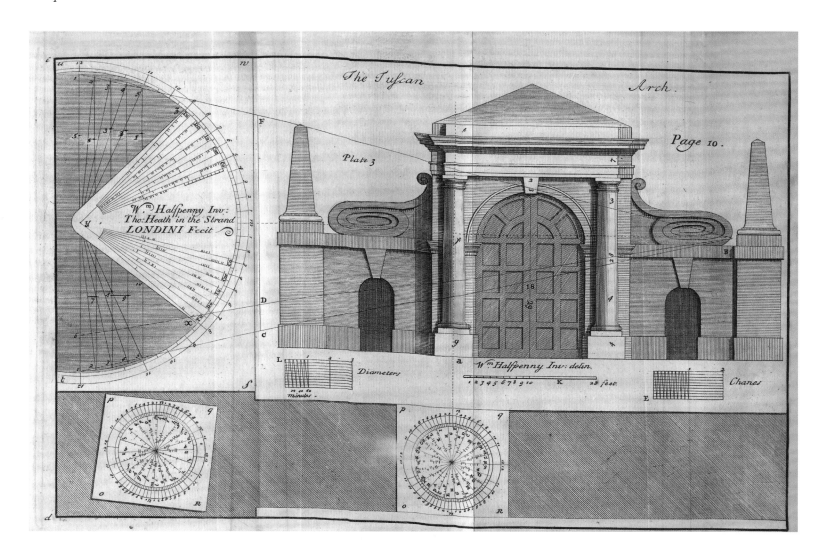

108. Fold-out plate showing specialized drawing board, from Halfpenny, *Magnum in Parvo* (1728)

book's title page, which gave Heath as the publisher along with John Wilcox, promised that the method was 'so plain & so easy, that a young Gentleman tho' an utter stranger to the Art, may apprehend the Whole, by seing only one Example wrought'. This appeal, however, appears to have been overstated. The rarity of the instruments, along with the single issue of the book, suggests that there was at best a limited market for the device. In contrast, Halfpenny had far greater success with his *Practical Architecture* (1724), which consisted of engraved tables of the orders with the dimensions of all the parts tabulated for a range of column diameters at half-inch intervals. The low cost of this pocket book would certainly have made it a more attractive proposition to workmen and beginners.

While Halfpenny's drawing board was a genuine novelty, Heath's next project as architectural publisher was for an English version of a century-old Italian instrument. Ottavio Revesi Bruti's *Archisesto per formar con facilità li cinque ordini d'Architettura* had been published in Vicenza in 1627. It described the operations of a sector whose legs moved over an arc engraved with the proportions of the orders and which included not just the parts relating to columns but also to arches, doors, and niches. Thomas Malie's straightforward English translation appeared, with a dedication to Lord Burlington, in 1737.[25] As the title-page vignette makes clear, the legs are positioned over the arc or limb and scaled dimensions taken off with compasses (figure 110).

Gentlemen, Practitioners, and Instrumental Architecture

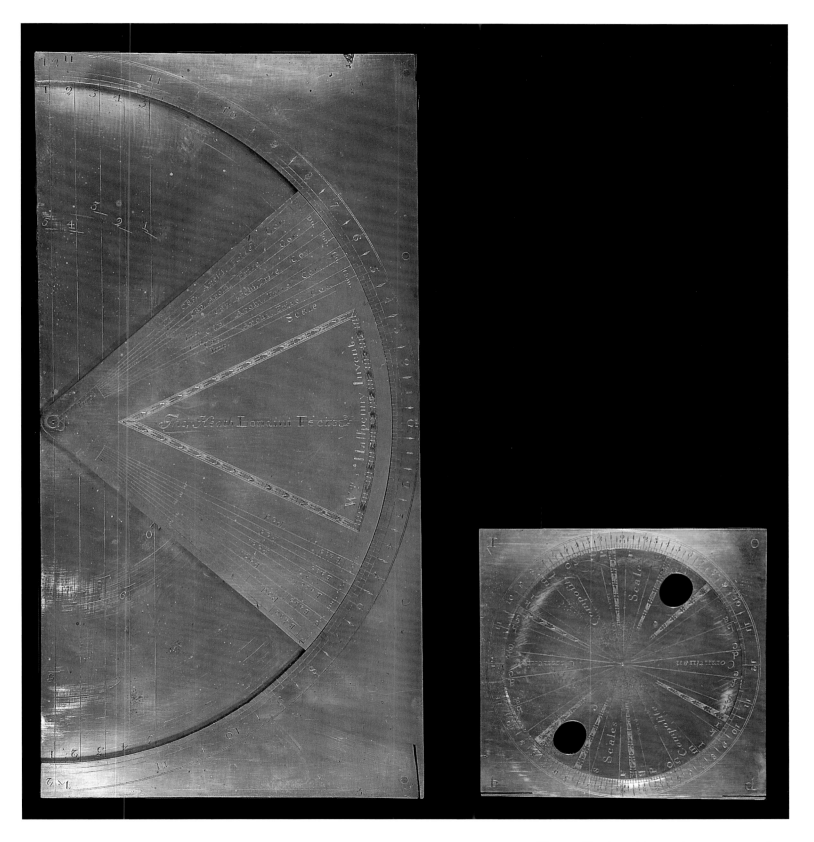

109. Thomas Heath, architectural protractors to the design of William Halfpenny, c.1728. Collection of Howard Dawes. Catalogue 58

125

110. Title page, from Malie, *A New and Accurate Method* (1737)

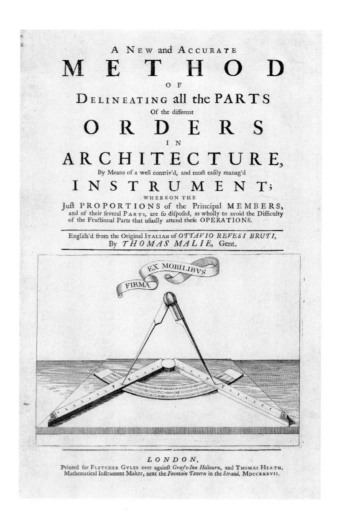

No examples of the Italian original are known – and the original treatise seems to have sunk almost without trace or comment – but an unsigned English example attributable to Heath survives (figure 111, catalogue 59). For readers unwilling to commit to the cost of a brass instrument, the book recommends cutting up the engraved plate of the parts to create a pasteboard version.

Malie's dedication acknowledged Burlington's 'careful perusal' of the text and his encouragement to publish, but it is not clear whether this should be considered a contribution to Burlington's Palladian programme. Revesi Bruti says nothing of his sources, and Malie added no further comment. Although Joshua Kirby would later claim that the instrument was based on Scamozzi, the proportions embodied in the limb of the instrument are not easily identified.[26] Perhaps it was enough for Malie and Burlington that the author had described himself as 'gentilhuomo Vicentino'. Alternatively, the instrument's appeal may have lain in its comprehensive character. Compared to Heath's other devices, it represented a much fuller realization of Roger North's ambition to embody the orders in a single compact device, thereby dispensing with the need for books and tables. Heath's role as maker and co-publisher indicates that, by the 1730s, he was considered the natural choice for such a project. He would surely have appreciated the running head throughout the volume: 'Instrumental Architecture'.

The last of Heath's ventures in this area was the *Treatise of Such Mathematical Instruments, As Are Usually Put into a Portable Case* (1747) by John Robertson, a Fellow of the Royal Society and author of texts on both navigation and mensuration. As the title indicates, the book was not

111. Thomas Heath, architectonic sector to the design of Ottavio Revesi Bruti, *c.*1737. Museum of the History of Science, Oxford. Catalogue 59

112. Folding frontispiece, from Robertson, *Treatise of Such Mathematical Instruments, As Are Usually Put into a Portable Case* (1747). Science Museum, London. Catalogue 60

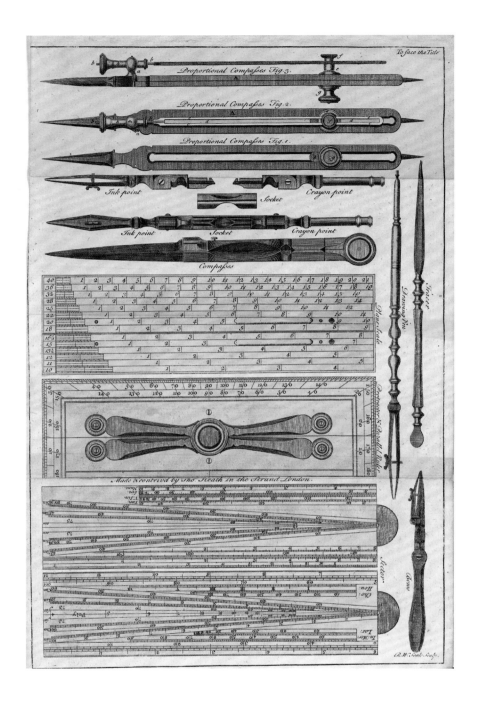

specifically devoted to architectural instruments but rather covered familiar tools for general technical drawing: compasses, pencils and pens, the parallel ruler, protractor, plain scale, sector, and proportional compasses (figure 112, catalogue 60). Architecture is nevertheless given a prominent place and its practitioners are the first to be addressed in the preface: 'The Architect, whether civil, military, or naval, never offers to effect any undertaking, before he has first made use of his rule and compasses; and fix'd upon a scheme or drawing, which unavoidable [*sic*] requires those instruments, and others equally necessary'. Robertson devoted considerable space to the orders in his discussion of the sector, even expanding his treatment in the second edition of 1757. The method of scaling, on the other hand, was unadventurous, utilizing only the most general line of equal parts in consultation with three tables of dimensions derived from Palladio.

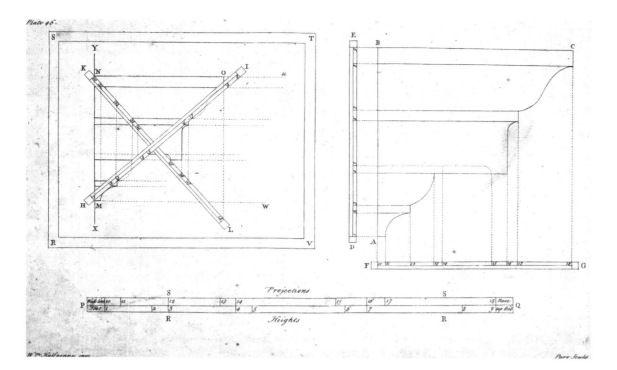

113. The 'Architectonick rule', from Halfpenny, *A New and Compleat System of Architecture* (1749)

This was, in other words, the same cumbersome technique that North, Carwitham, and other authors had tried to overcome and avoid.

What should be made of this final text? It does not describe a special-purpose architectural instrument, but rather tries to accommodate architecture to the draughtsman's standard equipment. In setting out the orders with an ordinary sector, Robertson accepts the 'inconvenience' of resorting to printed tables. Is this a retreat from the more ambitious instruments that Heath had previously published?

We can certainly find evidence elsewhere of pragmatic retrenchment. William Halfpenny, for example, returned to the orders in the second half of his publishing career, from 1748 until his death in 1755, now combining them with treatises on rural buildings as well as newly fashionable Chinese and Gothic garden architecture. Rather than the adapted and accessorized drawing board of his *Magnum in Parvo* – a device only available from Heath – he came up with a do-it-yourself method. In his *A New and Compleat System of Architecture* (1749), he announced a 'new invented architectonick rule … suitable to the proportions of any mouldings'. The rule consisted of nothing more than marking dimensions from a favoured authority onto a flat strip of wood and rotating it an appropriate amount to reduce in proportion for the required dimension. The principle is exactly that of the drawing board plates, but implemented in an inexpensive and ad hoc method (figure 113).[27]

As Halfpenny's last invention suggests, the tradition represented by Heath's architectural engagement appears to have died out. The latter years of the eighteenth century saw nothing comparable to his earlier innovations. Should Heath's support of a mathematical and instrumental approach to architecture therefore be deemed a failure? On an individual level, it is hard to imagine that his – doubtless limited – production of specialized architectural instruments was an enormous commercial success. Yet Heath's interest in this field may not have been primarily commercial: he appears to have had ample 'bread-and-butter' work. Architecture instead marked Heath out as a distinctive maker, not only in going beyond the routine production of his

114. Thomas Heath, set of drawing instruments, c.1740. Museum of the History of Science, Oxford. Catalogue 61

contemporaries and competitors but also in fashioning himself and his art as distinguished, ingenious, and polite. John Robertson, for one, lauded this maker, 'whose skill in contrivance, and care in executing the workmanship of curious instruments, [was] not, perhaps, to be surpassed by any artist'.[28] Even in a seemingly unspectacular portable case of instruments, Heath could make use of ingenious novelties such as a multipurpose pair of 'turn-up' compasses (figure 114, catalogue 61).

The fading of this instrumental enthusiasm, or its refocusing on a more limited and pragmatic approach, should not be taken as the peculiar fate of architecture. The creation and advocacy of 'universal' instruments and their rejection by working practitioners in favour of simpler and less ambitious alternatives is actually the typical pattern in most mathematical arts. It is precisely the story, for example, of English surveying in the sixteenth and seventeenth centuries, when authors gradually dropped their calls for the introduction and use of ambitiously exotic instruments.[29] Practical mathematics was not rejected by architects. Rather, there was a negotiated compromise in which an established repertoire of scales, sectors, protractors, and the other contents of portable cases became the minimum requirement for technical drawing. The universal, flamboyant or special-purpose creations might not have taken root but, as Robertson's text makes clear, the everyday products of the mathematical instrument maker were generally accepted as necessary for the art of drawing.

Chapter Eight
Raised High, Brought Low:
Architecture and Mathematics around George III

Two institutions founded during the reign of George III would redraw the disciplinary landscape of architectural practice. The first was the Royal Academy of Arts, created in 1768 after decades of scheming, dissension, and fruitless effort. There, architecture was placed alongside painting and sculpture as a fine art, an object of taste and refinement of a kind thought suitable for royal patronage (figure 115). The second institution was the Society of Civil Engineers, created in 1771, in which design was understood as a rational art of utility, strength, and economy for structures such as bridges, canals, and docks.[1] The shifts in sensibility that underlay these institutions had long been in motion. Between the academicians and the engineers, the middle ground on which architecture had traditionally stood was slipping away.

These institutional changes paralleled intellectual shifts. The ideal of beauty as objective and rooted in nature had underlain the emphasis on proportion, which Alberti took from Vitruvius and which was echoed by successive generations of architectural authorities. While the classical tradition had never been unchallenged, it was increasingly undermined by empiricist philosophy and by the articulation of new aesthetic doctrines. Most strikingly and directly, Edmund Burke attacked the idea of proportion as a guide to beauty in art. To him, proportion and quantity were 'a matter merely indifferent to the mind ... because there is nothing to interest the imagination', so that 'beauty is no idea belonging to mensuration; nor has it anything to do with calculation and geometry'.[2] This assault on the absolute and the assertion of the relativity of taste undercut the philosophical basis of the architectural canon. If mathematics could provide no objective warrant for beauty, judgement in matters of architecture would have to rely on no more than custom and convention. The painter Allan Ramsay dramatized this radical uncertainty in his *Dialogue on Taste*, when the freethinking colonel responded to the assured certainties of his noble interlocutor:

> LORD MODISH: But sure, Colonel, there are rules for the beauties of architecture, and not the smallest ornament of a base or cornish without its setled proportion.

115. C. Grignion after William Hogarth, frontispiece to the *Catalogue of Pictures ... &c. Exhibited by the Society of Artists*, May 1761

Compass and Rule

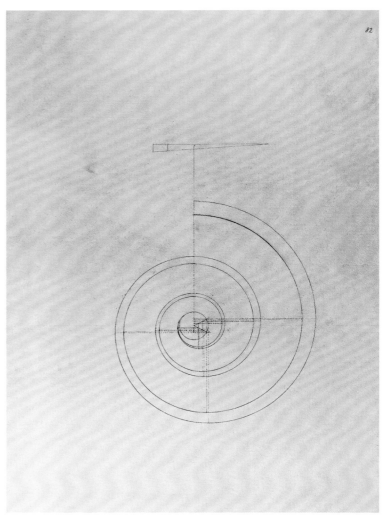

116. George III, study of the Ionic order, late 1750s. Royal Collection. Catalogue 62

117. George III, volute study, late 1750s. Royal Collection. Catalogue 63

COL. FREEMAN: … These rules are plainly no more than the analysis of certain things which custom has rendered agreeable. …. I should be exceedingly glad to hear a reason why an Ionic capital clapt upon its shaft upside-down should not become by custom as pleasing a spectacle as in the manner it commonly stands. I know this would be lookt upon as a sort of blasphemy by some of our delettanti; but so is every opinion, however reasonable, which opposes what is by custom established in any country.[3]

If the orders could be boldly overturned and proportion unseated what would be the fate of architecture? The answer, for the time being, was that custom prevailed. Palladian and, increasingly, neoclassical architecture continued to reign for public buildings and residences, but they did so without universal authority. Horace Walpole could refashion Gothic as a viable artistic category, while the vogue for Chinese buildings during the 1750s demonstrated the public appetite for architectural novelty drawn from outside the classical and European heritage.

The final section of our study examines architectural drawing and mathematics in this reshaped world, through the figure of George III. George was tutored in architecture before he came to the throne and remained keenly interested in and appreciative of the art throughout his reign. Just as Lord Burlington had demonstrated the vitality of the ideal of the gentleman-architect, George represented its culmination: who was more highly placed than the future king? He was also a patron of the mathematical tradition of architectural instruments. Within the context of the court, these objects took on new meanings, which reinforced their distance from professional practice.

This section also examines the roles of other key figures in the Prince's circle. John Stuart, Earl of Bute was the prince's tutor and his closest advisor. Rising briefly to the position of prime minister early in George's reign, he directed and encouraged the king's programme of political and cultural patronage. Bute appointed William Chambers as the prince's tutor for architecture and Joshua Kirby for perspective. Working with the royal instrument maker George Adams, Kirby oversaw the manufacture and publication of several new architectural instruments. In contrast to earlier devisers of instruments, his adherence to mathematics was not absolute, and his association with William Hogarth brought these objects into the turbulent world of satire. The tradition of elaborate instrumental architecture ends its career not as practical tool or polite accomplishment but within the arena of political and artistic conflict.

Given his royal status, George was hardly typical of the young men who took up posts as assistants and draughtsmen in the drawing offices of professional practices.[4] Yet it is because of his unique status that his student exercises still survive. The Royal Collection at Windsor Castle preserves roughly 170 architectural drawings, either wholly in the prince's hand or made jointly with his tutors. These range from basic exercises, produced in the late 1750s when the prince was still a teenager, to the polished, sophisticated proposals of an expert amateur, some made as late as 1777, when the king was almost forty years old. The drawings contain few virtuoso performances or extraordinarily imaginative designs. On the contrary, they are valuable precisely for their ordinary and sometimes pedestrian character. A beginner's first steps in the art would have normally been recycled or discarded. Few prospective architects had the luxury of practising on large clean sheets of fine paper, which were then preserved as evidence of youthful attainments. The corpus is thus one of the most complete sets of records that we have of architectural tuition in the period.[5]

The sequence of drawings displayed here has been chosen to represent the basic stages of the prince's architectural education. These proceed from the elementary tasks of constructing the orders, through composing plans and elevations, to contriving new inventions for basic building types, and finally to developing schemes for palaces and public buildings. During this process, we also see George learning attendant skills, such as the application of washes and the drawing of elements and buildings in perspective. Many of the prince's drawings, particularly the early studies, still display the evidence of their construction, namely the scoring, compass pricks, and guidelines produced by the operations of the dividers and ruler. From all appearances, the prince's tutors understood the study of architecture – at least at the elementary level – as a system of geometrical procedures, a protocol of proportional design rules. This was manual and instrumental work, but it was nevertheless considered appropriate – indeed essential – for a future monarch. In this respect, the collection represents the culminating moment of a centuries-old effort, for it demonstrates the extraordinary success of architects and theorists in promoting the idea of design as a socially and intellectually noble pursuit of both mind and hand.

A series of some twenty-two studies of the five orders represents the most likely starting point for the prince's studies. As Jane Roberts has observed, some of these are very close to those illustrated in Chambers's *Treatise on Civil Architecture* (1759).[6] Of this group, we include three

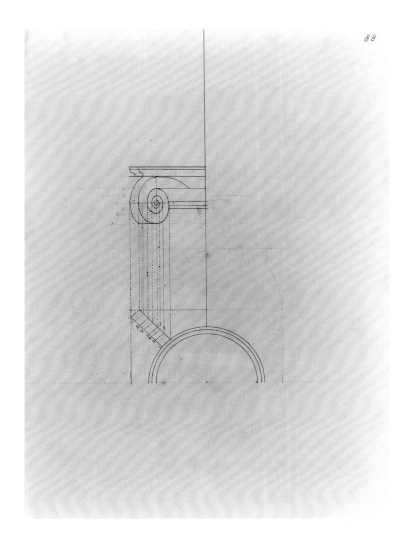

118. George III, combined plan and perspective study of a volute, late 1750s. Royal Collection. Catalogue 64

119. George III, drawing for a high tower of five stories, late 1750s. Royal Collection. Catalogue 66

related drawings of the Ionic order, which reveal the prince attending to a variety of graphic techniques. An overall view shows the relationships between entablature, base, and capital (figure 116, catalogue 62). As in most examples of the series, the handling of the wash lacks confidence, and the process of construction is clearly visible through the many guidelines and compass pricks. Many of the latter are concentrated in the area of the volute, a notoriously difficult form to master. One dictionary of architectural terms had defined the spiral as 'a Master-Piece of the Compasses'.[7] Chambers must have felt the exercise important, for he directed the prince to produce at least one detailed study of the volute, in what must have been a common preparatory exercise for architectural students of the time (figure 117, catalogue 63). The method is that of Nicolaus Goldmann, as described in Chambers's *Treatise*. George's pen trials are evident at the top margin of the page. A further study required him to render the volute in perspective (figure 118, catalogue 64); the view is derived from the plan in a clear and neat graphic operation.

From such elements, the prince was led on systematically through ever more complex problems of layout and composition. In a sketch of a palace elevation, we see a trace of the instructions that guided George's exercises (figure 120, catalogue 65). Endorsed on the reverse as 'His majesty's palace', the sketch is liberally supplied with dimensions and notes, some of which are

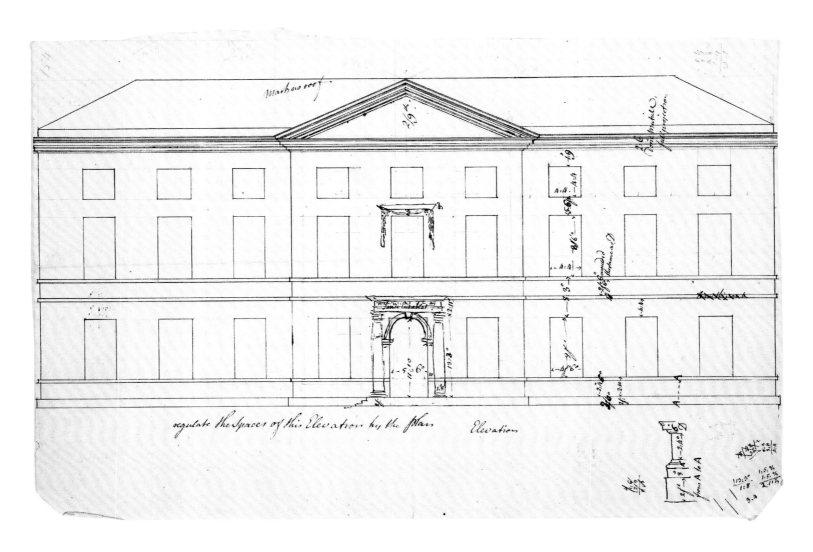

120. George III, sketch of palace elevation, with notes and observations, late 1750s. Royal Collection. Catalogue 65

simply labels ('Ionic Entablat[ure]', 'Doric'), while others record more of the drawing process ('mark no roof', 'Molded the same as D'). As David Watkin has noted, the drawing appears to be an exercise in which the prince was instructed to 'regulate the Spaces of this Elevation by the Plan', that is, to correct the alignment of the windows in the three central bays. Chambers also gave his royal pupil opportunities to engage in more inventive exploration and design. The collection contains several related series of drawings for towers, domed buildings, temple fronts, and pavilions. These studies, too, were probably exercises, permitting the student to experiment with the arrangement of familiar architectural elements. A drawing for a five-storey tower, for example, is likely to have been conceived as a folly, perhaps with a view to ornamenting the gardens at Kew (figure 119, catalogue 66). The composition is pitted with signs of its construction, with many centre points marking the use of the compasses and scribed arcs for setting off dimensions along the vertical centre-line, presumably taken from the scale bar along the bottom.[8]

The prince's more advanced work typically took the form of plans and elevations for palaces and grand public buildings. The collection contains many finished drawings of this type but, in some ways, they reveal less of George's progress as a student. Having been cleaned up to a presentation standard, they typically lack the traces of the design process that characterize his

121. George III, preliminary sketch of a palace façade, unknown date. Royal Collection. Catalogue 67

122. George III: perspective view of a palace with landscape, unknown date. Catalogue 68

123. Title page, from Kirby, *Perspective of Architecture* (1761). Collection of Howard Dawes. Catalogue 69

earliest studies. It is for this reason that a preliminary drawing of a palace façade stands out (figure 121, catalogue 67). The frontispiece, flanked by Serlian windows, carries a colossal order of pilasters. The guidelines are still evident, showing how the elements were set out and aligned. While he may have received some assistance with the drawing, the lightly pencilled-in Corinthian or Composite capitals and the very finely spaced entablature lines show the prince now in greater control of his draughtsmanship. Beyond elevations and plans, and the limited domain of the five orders, George was also led on to more developed perspective views. The labelled guide marks at the upper margin of a topographical study of a palace wing suggest that this was an exercise, presumably under the guidance of Joshua Kirby (figure 122, catalogue 68).

On his accession in 1760, George immediately implemented a policy of significant cultural and artistic patronage. In deliberate contrast to George II, the new king began awarding pensions to literary and scientific figures, while undertaking a major programme of collecting, with particular enthusiasms for books, maps, paintings and drawings, medals, clocks, and scientific instruments. Kirby's luxurious, large-format volume, *The Perspective of Architecture*, published in 1761, was intended to capitalize on these circumstances (figure 123, catalogue 69).[9]

The first part of Kirby's work – not found in all copies – consisted of 'The Description and Use of a new Instrument called the Architectonic Sector'. The second part, illustrated with a magnificent series of plates, contained 'A New Method of Drawing the Five Orders, Elegant Structures &c. in Perspective'. Despite the claim of the title, the sector was not an entirely new device but rather a revised version of the Revesi Bruti architectonic sector, made available in England in 1737. Kirby's variant, however, served the same purpose. The device made it

Raised High, Brought Low: Architecture and Mathematics around George III

possible to call up swiftly the many minute proportions of the five orders – including the parts of pedestals and entablatures – for columns of any given height. Those measurements could then be easily transferred to paper with a pair of dividers. In altering the earlier version, Kirby rearranged the engraved scales on the arc and also changed their proportional basis (figure 124, catalogue 70). He pointed out that whereas Revesi Bruti's instrument was founded on Scamozzi, for the new sector 'the orders are taken chiefly from Palladio, corrected however by the purest examples of antiquity'.[10]

George's active interest in the work is evident on a number of levels. He was, in the first place, the dedicatee and had evidently subsidized the publication. The title page proclaims that the work was 'begun by command of his present Majesty when Prince of Wales' and Kirby is identified as 'Designer in Perspective to his Majesty'. The sumptuous, not to say ostentatious, copperplate dedication to the king repeats the claim of the title page that the work had been composed at the king's request, adding that it was 'carried on under your EYE, and now Published by Your Royal Munificence'. George was, moreover, involved in the composition of the work itself: at least one of the plates was reproduced from a drawing by him.[11] In the Royal Archives at Windsor there is also a manuscript in George's hand of the first section of the text on the sector. It is not a copy of the printed work, but a preliminary draft with variants, deletions, and revisions that take the text nearer its final form.[12] The instrument must have appealed to the

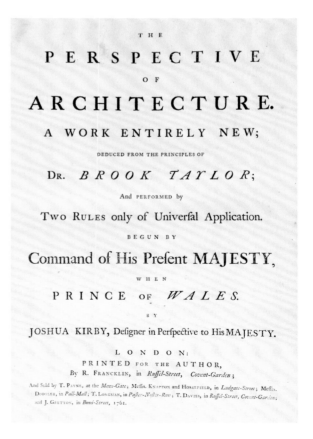

124. George Adams, architectonic sector to the design of Joshua Kirby, 1760s. Collection of Howard Dawes. Catalogue 70

125. George Adams, architectonic sector to the design of Joshua Kirby, 1757–60. Royal Institute of British Architects, London. Catalogue 71

Compass and Rule

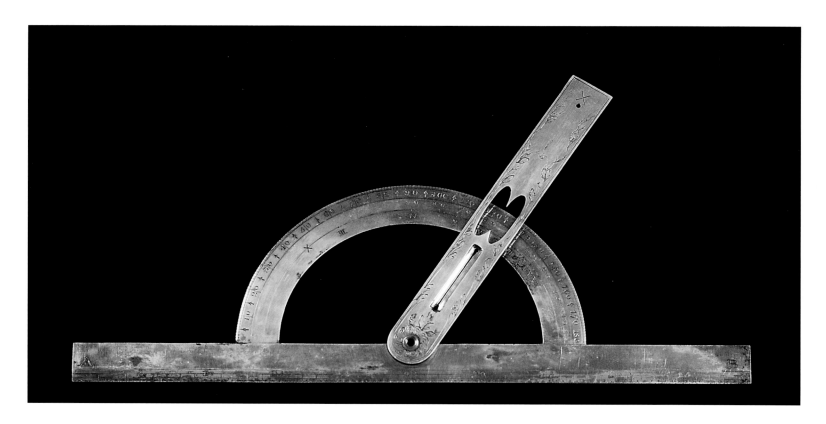

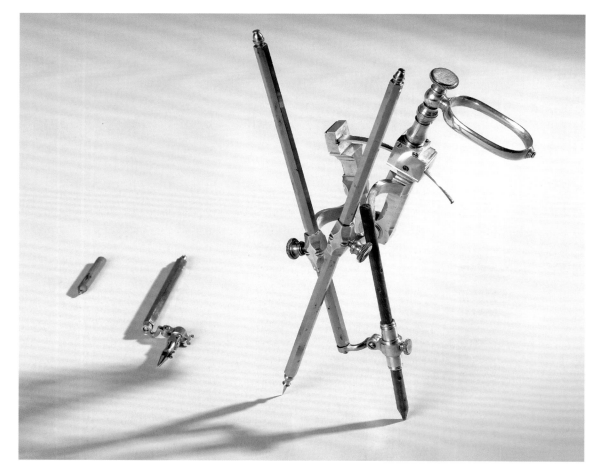

126. George Adams (attr.), protractor for Queen Charlotte, designed by Joshua Kirby, 1765. Museum of the History of Science, Oxford. Catalogue 72

127. George Adams (attr.), volute compass, designed by David Lyle, 1760. Science Museum, London. Catalogue 73

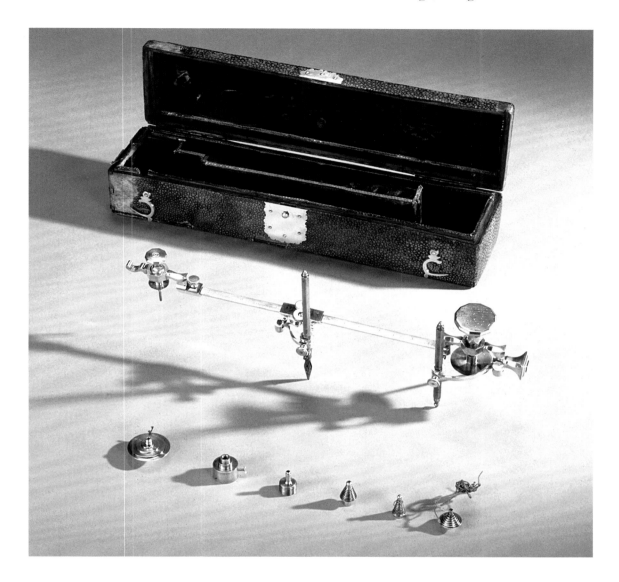

128. George Adams (attr.), volute compass, designed by David Lyle, 1760. Science Museum, London. Catalogue 74

king both for its practical value and for its aura of precision and mathematical ingenuity. Although evidence of its use in his own architectural drawing has not yet surfaced, he was undoubtedly well-acquainted with it, if only from lessons with his drawing master. As Kirby stated, the sector had been used to lay down the measures in the book.[13]

There are at least six extant examples, all by the royal instrument maker George Adams.[14] Compared to the rare survival of architectural instruments produced earlier in the century, this seems like flourishing success, but the circulation of Kirby's architectonic sectors appears to have been limited to the circle around the royal household. The most impressive survivor is a fine silver sector with the additional refinement of a retaining strut to preserve the opening of the sector's legs when in use (figure 125, catalogue 71). Just such an elegant and improved version would have been appropriate for a patron, perhaps the Earl of Bute, who is recorded as having owned several sectors, as well as many other scientific and mathematical instruments. Indeed, Bute was a collector on a huge scale – the auctions after his death ran for forty days in 1793 and 1794. Among his mathematical instruments was 'A silver architectonic sector, with apparatus, &c. by Adams, in a case, weight of silver about 24 oz'. Bute also owned an ivory architectonic sector, again by Adams, as well as two examples of an otherwise unattested perspective sector designed by Kirby.[15] Other court patrons also followed the vogue for drawing instruments: a silver protractor signed 'Joshua Kirby Invt. Octr. 5. 1765' carries the royal cipher of Queen Charlotte on the reverse, suggesting that Kirby's title of 'Designer in Perspective to Their Majesties' represented a genuine relationship with the queen as well as the king (figure 126, catalogue 72).[16]

The tight network of invention, manufacture, and patronage focused directly around the king is equally evident from a group of four so-called volute compasses. Designed to draw spirals of varying kinds, these unique instruments could be used for general geometrical exercises as well as the characteristic form of the Ionic capital. Three of the four are hand-held, all to the same basic design (figure 127, catalogue 73). The fourth is larger and carries a dedication by David Lyle, the designer of the set, that addresses George as king (figure 128, catalogue 74). Since it is dated 1760, the compass must have been presented soon after his accession, but it was not a speculative gift from an unknown aspirant. Lyle reveals the contacts and encouragement that he received in his *Art of Shorthand Improved* (1762). This work is dedicated to Bute, and Lyle recalls that 'By your Lordship's good offices, I was enabled to bring my new mathematical instruments to great perfection; and at your desire, I compleated a set of them for the use of his Majesty'.[17]

Bute had instilled in his royal pupil a sense of future responsibility for the encouragement and protection of learning and the arts, which sometimes makes his role as patron difficult to disentangle from that of George. Another outcome of their close personal and working relationship is found in an extraordinarily elaborate silver microscope created for the king (figure 129, catalogue 75). The microscope was signed, once again, by Adams, although the designer of the decoration is unknown. In its structure, the instrument contains a knowing nod to the king's tastes by its incorporation of a fluted Corinthian column as its main support. Two flanking bases ornamented with miniature vases enhance the architectural programme. Although the decorative extravagance might appear to interfere with its overt function, this was in fact a working device, incorporating both a compound and a simple microscope. It was not only used for the private instruction and entertainment of the king and his children but was also accessible to visitors and widely reported as a symbol of the king's enlightened tastes.[18]

Kirby's instruments did not differ markedly in conception from those advertised by Thomas Heath earlier in the century. A certain continuity is also evident in the instruments' manufacture. George Adams had been apprenticed to Heath and, like his entrepreneurial master, appears to have played an active role in devising and promoting his instruments, including Kirby's own sector.[19] Yet, for all this commonality of form and manufacture, the wider social and intellectual context had shifted. For Heath, architecture had been a way to give polish to mathematics and to extend its genteel reach. By the mid-century, however, the leading London makers no longer identified themselves as specialists in only mathematical instruments. They had become mathematical, optical, and philosophical instrument makers, with wider scientific and commercial interests and a larger range of merchandise. While practical mathematics was becoming more solidly professional, some sciences had become newly fashionable. Electricity, for example, offered spectacular opportunities for polite instruction and entertainment, and new devices such as the solar microscope could project the invisible world onto a large screen for an assembled audience. Kirby's courtly role remained prestigious, but it was not central to these public developments.

One outspoken English author on perspective, Thomas Malton, corroborates this impression. In a long and highly critical review of his English predecessors, he reported sarcastically that a

Raised High, Brought Low: Architecture and Mathematics around George III

129. George Adams, silver microscope for George III, *c.*1763. Museum of the History of Science, Oxford. Catalogue 75

143

Raised High, Brought Low: Architecture and Mathematics around George III

great work had been expected from the prince's tutor, 'when, lo! in 1761 appeared a Colossus indeed, of gigantic stature, truly worthy of its great Author'. In reality, Malton continued, Kirby's tome was 'a pompous nothing' and the discussion of the sector 'so much waste paper (for I am of opinion, that the Instrument was never made so much use of since)'.[20]

The world of graphic satire threw up even more emphatic evidence of the instrument's marginality. During the turbulent politics of the early 1760s, the sector would be closely linked with the royal circle and, in particular, with Kirby's patron and George's prime minister, Lord Bute. Deployed as a convenient emblem of his patronage, the instrument played a unique role in the political and artistic controversies of the time. The prime document here is an anonymous print from 1762, 'A Sett of Blocks for Hogarth's Wigs'. Ostensibly a response to William Hogarth's own 'Five Orders of Periwigs', it was also triggered by Hogarth's politically partisan print of 'The Times, Plate 1'.

The connection to the architectonic sector comes through the close relationship between Kirby and Hogarth. The two men had been linked publicly since 1754, when Kirby had dedicated his treatise *Dr. Brook Taylor's Method of Perspective Made Easy, Both in Theory and Practice* to the painter. Hogarth reciprocated by supplying the work's celebrated frontispiece of false perspective (figure 130).[21] The easy familiarity between the two is evident from the bantering tone of a further request by Kirby. In addition to the frontispiece, he hoped that Hogarth might contribute an empiricist account of perspective to subdue any readers with an overly zealous adherence to mathematics:

130. Etching by Luke Sullivan after William Hogarth for the frontispiece to Kirby, *Dr. Brook Taylor's Method of Perspective Made Easy* (1754). British Museum, 1868-0822-1604

131. William Hogarth, preparatory drawing in red chalk for the frontispiece to Kirby, *Perspective of Architecture* (1761). British Museum, 1860-0728-65

> If the little Witlings despise the Study of Perspective, I'll give 'em a Thrust with my Frontispiece which they cannot parry; & if there be any that are too tenacious of Mathematical Rules, I'll give them a Cross-buttock with your Dissertation, & crush 'em into as ill-shaped Figures as they would Draw by adhering too strictly to the Rules of Perspective.[22]

Hogarth was not persuaded, but he did later provide Kirby with a frontispiece for the *Perspective of Architecture* (figure 131). By now Serjeant-Painter to the King, Hogarth's design featured a newly composed architectural order in honour of George as Prince of Wales.

Hogarth presented architecture in more satirical vein in his 'The five orders of PERRIWIGS as they were worn at the late CORONATION, Measured Architectonically' (figure 132, catalogue 76). A classic example of Hogarth's wit, the print transfers the language and visual forms of the architectural treatise to a mock study of wigs. These objects are presented in five 'orders', from 'Episcopal or Parsonic' to 'Queerinthian', in shapes that range from the sexually suggestive to parodies of architectural analysis. One diagram, for example, is labelled 'Helices or Volute or Spiral or Curl'. While ridiculing the pompous and absurd fashions of both the established and the fashionable, Hogarth was also laughing at architecture itself, particularly at what he saw as its slavish devotion to antiquity. The advertisement beneath the image takes a well-aimed swipe at the recently published prospectus for James Stuart and Nicholas Revett's *The Antiquities of Athens, Measured and Delineated* (1762). The wickedest humour is directed at 'Athenian' Stuart himself. He is shown as a blockhead – literally a wooden head for hats or wigs, but a term that had long carried the connotation of utter stupidity. Compasses spring from his head and he is measured by a scale captioned 'One Nodule / 3 Nasos / each Naso 3 Minutes'. Stuart's 'naso' has been chopped off to disguise the resemblance, but this is too ingeniously light-hearted to be truly vicious, and apparently Stuart took the caricature with good grace.[23]

The response to one of Hogarth's subsequent prints, however, was not as mild. Indeed, Plate 1 of 'The Times', issued a year later, provoked a ferocious reaction (figure 133). Hogarth for once stepped into the arena of party politics here, endorsing the efforts of the new king and his prime minister, Bute, to negotiate a peaceful end to the Seven Years' War. At the centre of the scene is a fire engine being used to dampen the blaze of global war, the hose handled by a figure wearing the king's badge, 'G R'. Bute's Scottish background and entourage are alluded to in the foreground, with a kilted figure bringing water for the pump. Against these peaceful efforts are the opposition figures in the surrounding buildings, aiming their hoses at the firefighter rather than the fire. Meanwhile Bute's opponent Pitt the Elder, on stilts and weighed down by the millstone of his £3000 pension, fans the flames with bellows.[24]

For once Hogarth misjudged the public mood. He was assaulted by a stream of prints orchestrated by John Wilkes and his notorious opposition periodical, *The North Briton*. As Serjeant-Painter, Hogarth was depicted as a hired hand, cravenly dependent on Bute and creating propaganda in exchange for his position. But the odium directed at Hogarth was as nothing compared to the sustained and vitriolic campaign against the unpopular Bute. He was the object of a veritable explosion of political satire in 1762 and 1763, running to literally hundreds of prints. Strongly emblematic in character, these prints often scandalously associated Bute with the petticoat of Augusta, the Dowager Princess of Wales, and they invariably identified him by a jackboot. This was not only a pun on his name (James Stuart, Earl of Bute as Jack Bute) but apparently the origin of the term's metaphorical sense as tyrannical and oppressive.[25]

132. William Hogarth, 'The five orders of PERRIWIGS as they were worn at the late CORONATION, Measured Architectonically', third state, 1761. Collection of Howard Dawes. Catalogue 76

Raised High, Brought Low: Architecture and Mathematics around George III

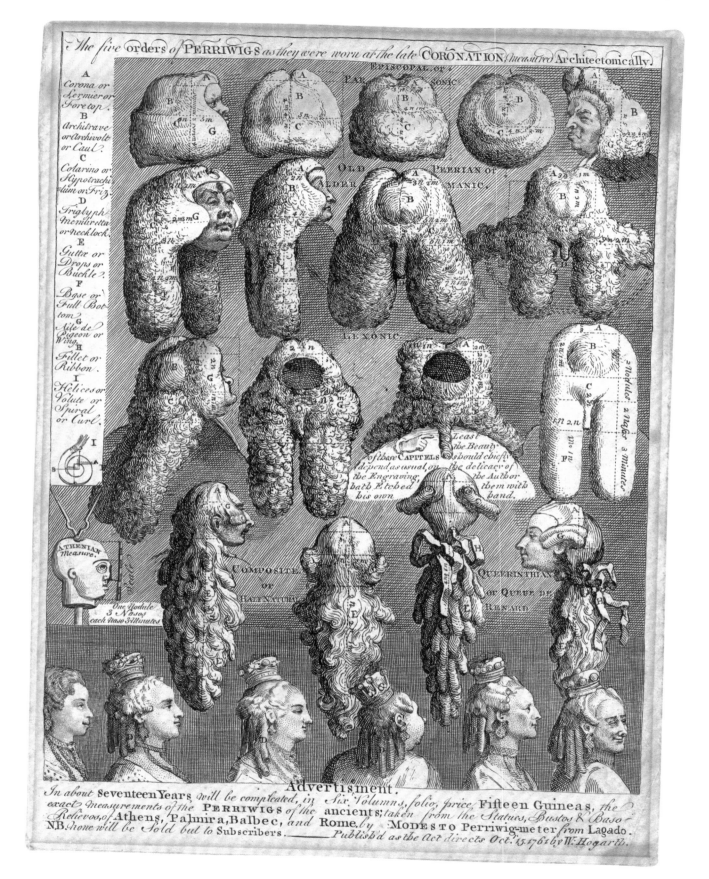

Compass and Rule

133. William Hogarth, 'The Times', plate 1, 1762. British Museum, Cc-1-172

This, then, is the context for 'A Sett of Blocks for Hogarth's Wigs' (figure 134, catalogue 77). The central figure is Bute, represented by his jackboot and with his Scottish origins indicated by the suggestively placed bagpipes. He is surrounded by an array of other pro-government blockheads and the print overflows with richly detailed references to the contemporary political turbulence occasioned by the government policy of peace. But, unique in the vast tide of visual denunciation facing him, Bute also appears here with a distinctive emblem: the unmistakeable form of the architectonic sector. The prime minister is satirically identified with an instrument of mathematics – though, rather than architectural scales, the characteristic arc here carries domestic animals. The anonymous author wanted to hit precise targets and make explicit connections. On the right are two additional blockheads, labelled K and H for Kirby and Hogarth, clearly identified as supporters of Bute. Kirby's connection is via the sector, slung under

Raised High, Brought Low: Architecture and Mathematics around George III

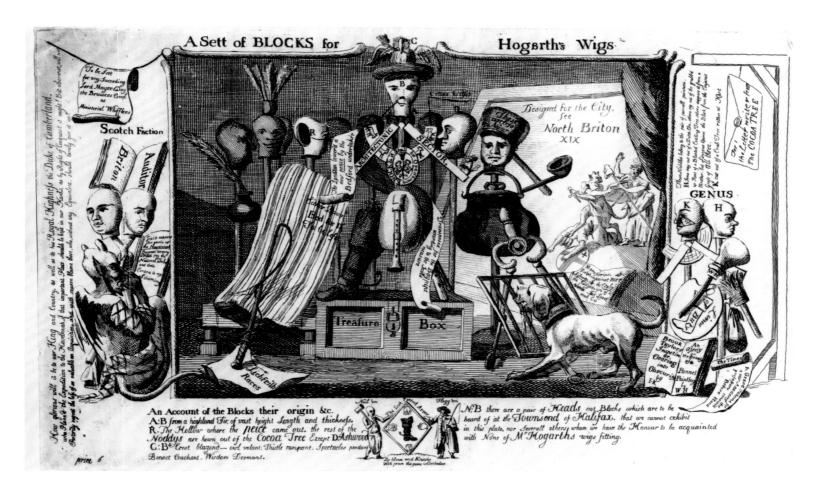

134. Paul Sandby (?), 'A Sett of Blocks for Hogarth's Wigs', third state, 1762. Gainsborough's House, Sudbury. Catalogue 77

his head as an identifying emblem, while Hogarth's palette carries not the famous serpentine 'Line of Beauty' from his *Analysis of Beauty* (1753) but a jackboot captioned 'Line of Buty'.[26]

Mathematics was here transformed from a polite accomplishment to a convenient badge. Used to pinpoint the enemy, the identification could only work because the architectonic sector was seen as the exclusive possession of a handful of elevated individuals. If it had been enthusiastically adopted by a broad audience, it would have had no emblematic purchase. Only because it had not moved out beyond the court circle of George, Bute, and Kirby could it play a satirical role. George's architectural enthusiasm had raised the sector to the heights of royal patronage and monumental publication. The price for this grandeur was its exposure to ridicule and its ultimate isolation as an irrelevance to the daily experience of working architects.

Another of Hogarth's friends brought the distanced eye of the outsider to the appraisal of the English cultural landscape. Jean André Rouquet's account of *The Present State of the Arts in England* appeared in 1755 and offered a sympathetic examination of English art and craft, from painting to furniture. Rouquet used psychological stereotypes to frame his account of individual arts and to seal the division of culture into separate spheres.

> 'Men of a bright and lively imagination, are unfit for a long series of calculations … On the other hand, persons of more judgment than imagination, are as unfit for the polite arts: with a rule and compass always in hand; in every thing that surrounds them, they require

Compass and Rule

135, 136. George Townshend (?), 'A scene of a pantamime [sic] entertainment lately exhibited. London: sold in May's Buildings, Covent Garden', November 1768. Museum of the History of Science, Oxford. Catalogue 78

the test of strict demonstration. This geometrical rigour, so agreeable to the nature of commerce, is a damp to genius, and removes the happy delirium which ought to preside over works of imagination. It was therefore natural, that the English should busy themselves chiefly in geometry, in mechanics, and commerce; while the French and Italians found their principal amusement in the polite arts'.[27]

Opposed to genius and happy delirium, mathematics was here not only condemned to narrowness but deliberately tainted by association with commerce. By this account, when architecture and perspective were placed among the polite arts, they fell outside the reach of rule and compass. Ironically, this was the ambivalent intellectual and aesthetic environment into which Kirby inserted his architectonic sector. In seeking Hogarth's aid for an empiricist defence of perspective, he acknowledged the senses rather than mathematical reason as the ultimate arbiter of the art. When the allegiance of even its author lay outside mathematics, it is not surprising that the sector's future was limited.

The fading star of mathematical practice, and of the instruments that served as its most tangible asset, is nicely captured by a last print. 'A scene of a pantamime [sic] entertainment lately exhibited' dates from 1768 and dramatizes the break-up of the Incorporated Society of Artists and the formation of the Royal Academy (figures 135, 136, catalogue 78). The warring factions are represented on stage, with the breakaway group as an eight-headed hydra fought by loyal members of the Incorporated Society. We are interested in a shadowy detail of the background. The painted backdrop features two figures in the clouds. On the left is Castor, clearly Sir William Chambers, overlooking an Academy constructed in the Chinese manner. His compasses are poised over a roll of diagrams or plans. The sketchily delineated instrument brandished in his left hand is once again the architectonic sector, but now scarcely distinguishable in the gloom and far distant from the foreground action. It will not be heard of again.

PART II

previous page:
137. Nicholas Hawksmoor, South elevation of the 'Revised design' for St Paul's, c.1686–87, All Souls College, Oxford II.29 (Geraghty 81)

138. Detail of the completed dome and crossing from the longitudinal section looking south, in Poley, *St Paul's Cathedral* (1927)

Geometry and Structure in the Dome of St Paul's Cathedral

GORDON HIGGOTT

Of all the drawings to have come down to us from the hand of Sir Christopher Wren, none is more revealing of his conviction that the 'geometrical ... is the most essential part of architecture' than his study design for the dome of St Paul's Cathedral in the Prints and Drawings collection of the British Museum (figure 86, catalogue 45).[1] Inscribed 'Sr Chr: Wren's owne hand' in the late handwriting of Nicholas Hawksmoor (probably after Wren's death in 1723), the drawing is in the loose freehand pen technique of the architect's latest known drawings and can be dated to about 1690, when construction had reached the internal attic above the Corinthian pilasters, immediately below the eight crossing arches (figures 138, 139), and work had begun on the design of the dome.[2] It is among very few drawings by Wren himself after Hawksmoor had joined his office as a clerk and draughtsman in the early 1680s, and is the only surviving study in his hand for the dome of St Paul's in the final stage of its design.

Especially significant is the geometric scheme that underlies the section, for it reveals Wren applying a mathematical formula for a 'cubic parabola' to define the curve of the middle dome and the projection of the peristyle and lower drum. Wren owed this formula to his scientist-colleague Robert Hooke, who had stated to the Royal Society in 1671 that the shape for a perfect dome was the 'cubico-parabolical conoid': that is, the dome formed by rotating the cubic parabola, $y = x^3$, about its y-axis (figure 140b).[3] Hooke had arrived at this formula from his work on the 'hanging chain': the catenary-shaped line of a weighted chain, which, when inverted, gave the equivalent curve for a masonry arch. The approximate mathematics of this curve were then thought to be $y = x^2$ (figure 140a), and Hooke's innovation was to use the 'cubic' version of this equation to give the corresponding curve for a dome. Historians have long speculated on the influence of Hooke's hanging-chain principle on Wren's designs for the dome of St Paul's.[4] This drawing is our only evidence for Wren's application of that principle, and its earliest known use before Giovanni Poleni derived his conclusions on the structural condition of St Peter's in 1748 from an experimental use of the hanging chain to determine the line of thrust within segments of its dome (figure 141).[5] Careful analysis yields the place of Wren's design in the sequence of his studies for the dome of St Paul's, and the relationship between its geometry and the fabric as built.

The drawing can be dated to a few years after a major revision to the design in 1685–87, prompted in June 1685 by a decision by James II's new parliament to treble the annual funding to St Paul's from the tax on coal coming into London.[6] Work had begun at the east end in June 1675, and, during the first decade of construction, Wren had intended to crown his

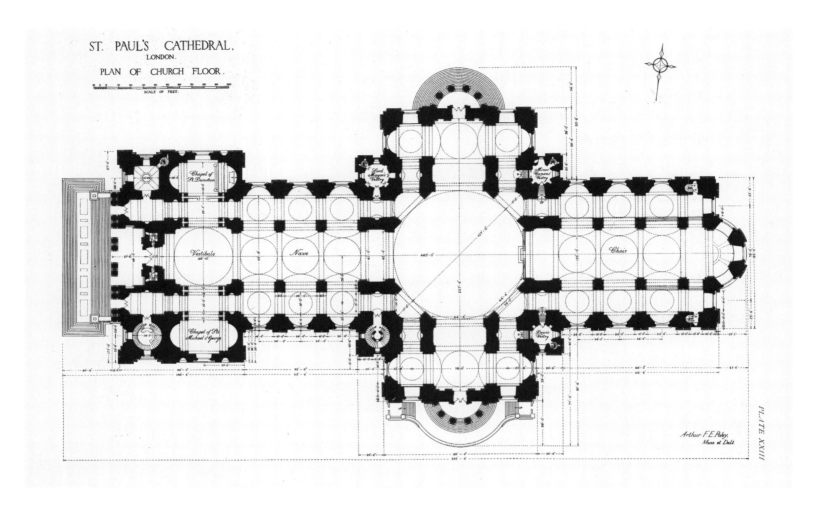

139. Floor-level plan, from Poley, *St Paul's Cathedral* (1927)

140. Comparative graphs of a catenary curve, $y = x^2$ (a), and a cubic parabola, $y = x^3$ (b), using the formula in Wren's study design (catalogue 45), where single units on the x-axis correspond to units of 8s on the y-axis

141. Demonstration of the correspondence between an appropriately weighted chain and a section of the dome of St Peter's, from Poleni, *Memorie istoriche della gran cupola del Tempio Vaticano* (1748)

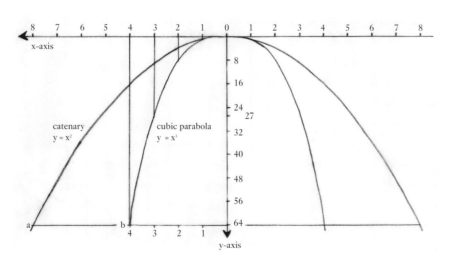

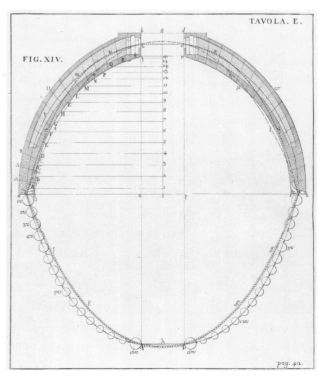

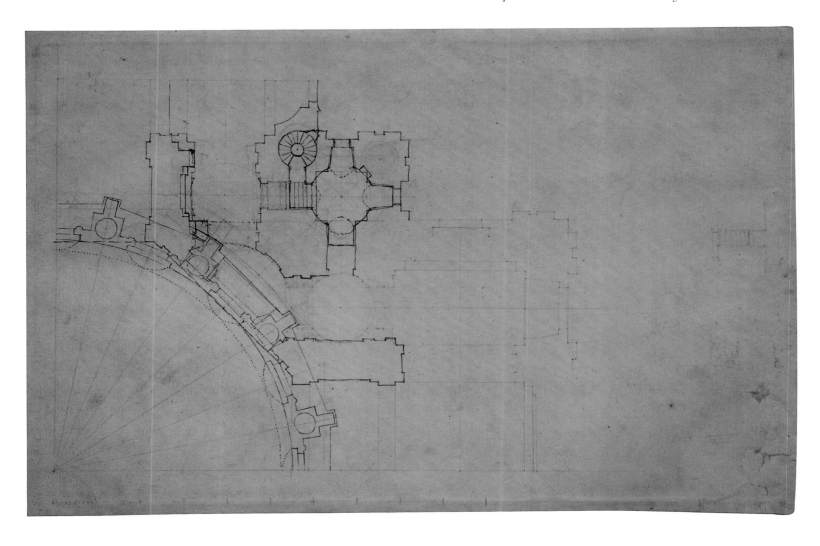

cathedral with a dome of modest height and bulk, in an incomplete design that has since become known as the 'Penultimate'.[7] The dome in this design would have had a sixteen-bay peristyle, only nine feet thick (figure 142), and an octagonal or circular lower drum extending down in the diagonal bays to the semi-domes above the aisles, allowing light to enter the central space through large windows in the drum at triorium level (figure 143). Preparatory studies indicate that the missing outer dome in this design would have been constructed in lead-clad timber, like the outer dome and lantern of the Warrant design for the cathedral, authorized on 14 May 1675 but immediately superseded.[8] However, in his Revised design of c.1685–87, Wren extended the plan of the cathedral, on which work had begun, and greatly enlarged his dome (figures 137, 145). He added a two-storey western body beyond the third bay of the nave to accommodate a library at the upper level, and built up his aisle walls to two storeys all round the cathedral, by replacing the balustraded parapet of his original design (on the left side of figure 143) with full-height 'screen walls' (figures 137, 154). He now proposed to crown the cathedral with a vast, richly modelled dome, built entirely in masonry and inspired, in part, by what he then knew of Jules Hardouin-Mansart's designs for a new royal church at Les Invalides in Paris, begun in 1677 and published in a suite of engravings in 1683 (figure 144).[9]

142. Christopher Wren, Quarter-plan of the dome and crossing, from the 'Penultimate design' for St Paul's, 1675. London Metropolitan Archives, SP 14 (Downes 5)

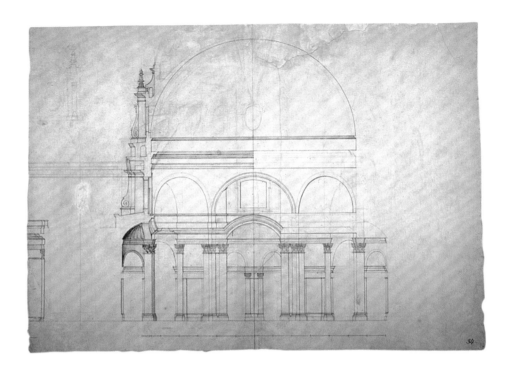

143. Christopher Wren, assisted by Edward Woodroofe, Diagonal section through the crossing and lower part of the dome, from the 'Penultimate design' for St Paul's, 1675, All Souls College, Oxford, II.34 (Geraghty 77)

144. View of the church of Les Invalides from the south, engraved by Jean Marot, from Boulencourt, *Description générale de l'Hostel royale des Invalides* (1683)

145. Proof engraving of east–west section, c.1687–88, London Metropolitan Archives, SP 118 (Downes 94)

The structural form of the dome of the Revised design is known from a proof engraving of the east–west section, prepared under Hawksmoor's supervision and datable to near the end of James II's reign (figure 145).[10] Its design depends partly on the dome of the Great Model (figure 146)[11] – the ambitious scheme for a centrally planned cathedral that Wren had reluctantly abandoned in 1674 – and partly on engravings of the two most important continental exemplars: a perspective cutaway view of St Peter's in Rome by Alessandro Specchi

Geometry and Structure in the Dome of St Paul's Cathedral

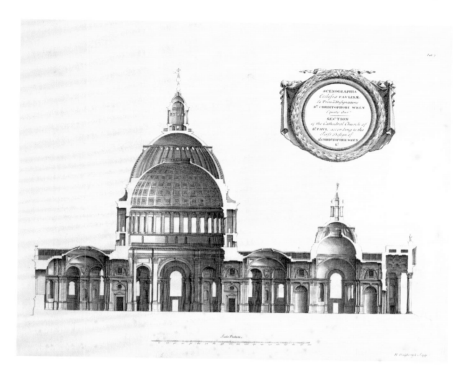

146. Section of the dome of the Great Model, engraved by Hulsbergh, 1726, from Wren, *Parentalia*, Sir John Soane's Museum, Drawer 69/2

147. Alessandro Specchi, Cutaway view of St Peter's, 1687, Sir John Soane's Museum, Drawer 59/7/2

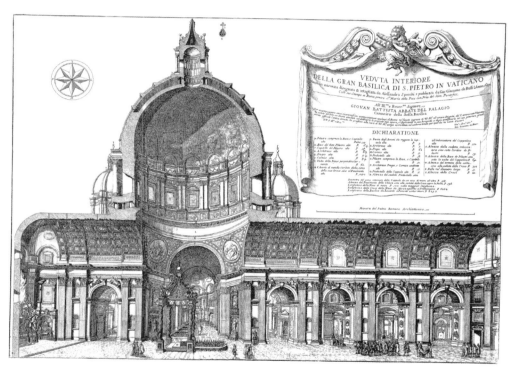

(figure 147),[12] and a section (on the diagonal axis) of the dome of Jules Hardouin-Mansart's Invalides church by Pierre Lepautre (figure 148), both issued in 1687.[13] Wren adopted the melon-shaped profile of the outer dome of his Great Model, but gave it a doubled shell to support a taller lantern, as at St Peter's, and borrowed from Hardouin-Mansart the device of lighting the soffit of the inner dome from attic windows concealed from view by a cove-shaped, truncated lower dome.

Geometry and Structure in the Dome of St Paul's Cathedral

The next design for the dome is probably the immediate precursor to Wren's study, and in its elevation it anticipates the profile of the dome that was eventually built (figure 149). This is a neatly drawn half-section and half-elevation of a modified version of the dome of the Revised design, drawn and annotated by Hawksmoor in a hand consistent with that in his many drawings for Hampton Court Palace in 1689–91.[14] Here, Wren refined the concealed lighting motif, but then changed his mind about the steeply curved outer dome in the half-section, and in two pasted additions reverted to a hemispherical outer dome, crowned by a tall lantern. It is clear, however, that the weight of this dome would have resulted in a massive load on the crossing piers. These piers had been built with cores of compacted rubble rather than squared masonry, and they rest on the rubble cores of larger piers at crypt level (figure 150).[15] Settlement started to occur in the piers of the crypt in the late 1680s and Wren must soon have realized that an all-masonry dome of the height and girth of that in figures 137 and 149 would have been far too heavy for the substructure at crypt level.[16] A related problem was how to light the interior of the dome effectively while maintaining its structural integrity.

Wren's study design in figure 86 is an initial attempt to reduce the mass of this all-masonry structure by thinning the shells of the inner and outer domes, reducing the size of the lantern, and introducing a parabola-shaped middle dome to determine the depth of the abutment and

149. Nicholas Hawksmoor, Part-section, part-elevation of a dome modified from the 'Revised design' of St Paul's, c.1690, London Metropolitan Archives, SP 171 (Downes 95)

150. W. Godfrey Allen, Isometric drawing of pier and superstructure, 1924, St Paul's Cathedral archives, A 15A

148. Diagonal section of the dome of Les Invalides, engraved by Pierre Lepautre, 1687, Bibliothèque nationale, Paris, (Va. 444, formatt 6)

to help transmit loads to the crossing piers. His design is for a twenty-four-sided dome, its peristyle in eight triads of openings, arranged so that each central arched window coincides with a crossing arch below, while the pilasters between them stand above the crossing piers. It can be seen from the more developed left-hand design that this middle dome offers little direct support to the small lantern. Instead, it supports the outer dome further down its slope, at a point where the outer and middle domes would help counteract the thrust of the inner dome. In this way the parabolic middle dome gives a common line of thrust for the structure as a whole. The middle dome also provides a route for a staircase to the lantern. This travels up the wall of the inner drum, over the back of the inner dome, through the middle dome and over its top to a spiral staircase beneath the lantern. It indicates that Wren drew the section on the diagonal axis – as he did for the 'Penultimate' design (figure 143) – since the stair would have reached the dome from the south-west corner bastion. The stairway is illuminated by small apertures in the outer dome and by borrowed light from the oculus of the inner dome. Unlike the oculus of the Great Model, however, this aperture would not have admitted light from above (figure 146); Wren therefore needed to find ways of improving the illumination in the interior. The line taken by the staircase, the use of decorative panelling on the soffit of the inner dome, and the tall proportions of the columnar drum, all suggest that, when he prepared this design, Wren made use of Lepautre's engraving of the Invalides dome on the diagonal axis (figure 148). But, while the superstructure of the Invalides dome was to be in lead-covered timber, Wren was proposing to build his dome entirely in masonry.

To ensure the structural stability of his dome Wren had to direct as much as possible of the load down to the broader and more solid front portions of the crossing arches, where deep soffit bands, or 'ribs', six feet across, in two, stepped courses of four feet and two feet, run down to the pilasters and half-pilasters that flank the piers.[17] The ashlar construction of the soffit bands and the projecting pilasters is illustrated in W. Godfrey Allen's isometric drawing, prepared in 1924 for a wooden model of a crossing pier, shortly before a major programme of structural reinforcement of the dome and its supports (figure 150).[18] The relationship between the wall of the inner drum, the soffits of the crossing arches, and the pilasters on the piers is clearly expressed in William Dickinson's cross-section of c.1696–1702, which records the dome as built up to the lower half of the peristyle (figure 88, catalogue 47).

In his study design (figure 86), Wren set his triple-dome structure on an inner drum that is designed to rest above these six-foot soffit bands. He then determined the depth of the abutment to his triple dome, beyond the inner drum, by means of the profile of a parabola whose base is the entire 160-foot square of the crossing area and whose upper part defines the middle of the three domes. Measuring on the cardinal axes, the 160-foot crossing square consists of the 108-foot diameter of the octagon and two 26-foot-deep crossing arches (figures 138, 139). This 160-foot dimension can be subdivided into eight 20-foot units, six of which (120 feet) give the external diameter of the inner drum, and two the 20-foot rear portions of the four main crossing arches, behind the 6-foot soffit bands.[19] Wren began on the right-hand side of the sheet by subdividing the half-section into four of these units with vertical lines. It is apparent from the scaled design on the left side of the sheet that these represent 20-foot units in the half-section, and that the third unit includes the 6-foot-deep wall of the inner drum. The external diameter of the inner drum in Wren's drawing is therefore 120 feet, as in his diagonal section for the 'Penultimate' design (figure 143), and in the fabric as built (figures 138, 88). He then drew the parabolic outline of the middle dome on an inverted graph whose 'y-axis' is the vertical centre-line of the dome. Here, the scale of numbers from top to bottom is not in feet but in units of eight, for ease of setting out the cubes of the divisions along the horizontal or 'x-axis' at the crown of the middle dome. These divisions are marked with dots along the baseline of the section, and vertical pen lines are drawn upwards to the x-axis, where they are 'stretched' outwards so as to equal the size of the units along the y-axis.

The cubic parabola is the curve generated on a graph by plotting the intersections of units along the horizontal or 'x-axis' with the cubes of those numbers derived from units on the vertical or 'y-axis' (figure 140b). Thus, in Wren's study, the line at the crown of the middle dome begins horizontally at the y-axis, since $0^3 = 0$; it meets the first vertical line at 1 on the scale ($1^3 = 1$), the second line at 8, or 2^3, the third at 27, or 3^3 (marked with a dot below 24 on the vertical scale), and it continues down to meet the baseline at 64 in a steeply curved line that defines the outer limits of the structure. Significantly, the top of the 'buttress' to the left of the inner wall aligns with the third intersection, at 27 on the vertical scale. The dome as a whole appears to be considered, both structurally and proportionally, as one part 'abutment' and three parts 'arch', and the overall height of the half-section of the middle dome is eight of these parts, making the frame of the dome in the right-hand sketch a perfect cube above its base at the crown of the crossing arches.

On the left side of the sheet Wren redrew this geometrically inspired section over a neat grid of ruled pencil lines, giving twenty feet to each of the four subdivisions and marking a scale in five-foot units at the bottom of the vertical centre-line of his dome. He now set the base of his grid one unit, or twenty feet, below the entablature of the dome and adjusted the widths of his horizontal units to correspond exactly with those on the vertical scale. The cube proportion of the middle dome was disrupted, and the effect was to narrow the buttress slightly and reduce the overall mass of the dome. Nevertheless, Wren maintained much of the geometric coherence of his initial design and carefully distinguished between the six-foot-wide wall of the inner drum and the peristyle behind. The height of the parabola of the middle dome above the baseline of the entablature is nearly the same as the 160-foot crossing square (since its profile at this level does not reach the edge of the grid); the inner dome, like the outer dome, is still hemispherical, and its apex remains one unit, or twenty feet, below the middle dome. The top of this 'buttress' is still at the intersection with the third twenty-foot unit of the section but it has been redrawn as a peristyle resting on a lower drum, which steps out below the roofline as far as the parabolic profile of the inner dome.

Wren's use of the cubic parabola has its origins in a presentation made by Robert Hooke at the Royal Society on 7 December 1671 about a principle for the 'figure of the arch of a cupola':

> Mr. Hooke produced the representation of the figure of the arch of a cupola for the sustaining such and such determinate weights, and found it to be a cubico-parabolical conoid; adding, that by this figure might be determined all the difficulties in architecture about arches and butments. He was desired to bring in the demonstration and description of it in writing to be registered.[20]

There is no record of Hooke having supplied the 'demonstration' of his 'figure', and the only other information that we have about his presentation is from John Evelyn, who wrote in his diary for that day: 'In the afternoone at the R: Society were examin[e]d some draughts of arches to sustaine a Cupola'.[21] From this remark, and from Hooke's expression 'cubico-parabolical conoid', we can infer that the drawings of arches illustrated a dome whose profile was a cubic parabola. As noted above, this was Hooke's three-dimensional rendering of his principle of the 'hanging chain' (figure 141). He had found that the statics of a hanging cord in tension were the same as those of an arch in compression, and in 1675 published his discovery in anagram form: 'As hangs the flexible line, so but inverted will stand the rigid arch'.[22] His assertion, in December 1671, that his 'figure' could be used to resolve all 'the difficulties in architecture about arches and butments' suggests that he was applying his 'hanging chain' idea to dome construction, while also asserting its broader application as a structural principle. That he freely offered such advice to Wren is known from an entry in his diary for Saturday 5 June 1675, which concerns the design of St Paul's Cathedral: 'At Sir Chr: Wren … He was making up my principle about arches and altered his module [design] by it'.[23] However, neither Hooke nor Wren left anything in writing on the mathematics of the catenary, or on the application of the hanging-chain principle to the design of an arch and its supports, let alone to the design of a dome. It was not until 1697 that the Scottish mathematician David Gregory published an early solution to the equation of the 'catenaria' and a description of the structural application of the catenary shape. Here he made the important assertion that if the shape of an inverted chain lies within the masonry of an arch, then the arch will stand: 'none but the catenaria is the figure of a true legitimate arch, or fornix. And when an arch of any other figure is supported, it is because in its thickness some catenaria is included'.[24] Wren's study design shows that he too knew that the 'figure', or profile, of the arch was the key to determining the size of the abutment, even if its mathematics were only approximate, and that for an arch or dome to stand firm, its line of thrust had to be contained within the boundaries of its abutment. His design is the earliest known architectural drawing to express this principle, and it embodies the structural solution for his completed dome.

Wren modified his design for the dome several times in the years leading up to the start of work on the lower drum in early 1696, and again in about 1702 to resolve the structural form above the peristyle (figure 88, catalogue 47).[25] At the end of the first phase of the design, he devised the tapering inner drum, and at the beginning of the second, the hidden, flat-sided cone that supports the lantern. These two cone-like structures owe their inspiration to the parabolic dome in his study design, but neither is a pure geometric form. Wren arrived at their profiles, buttressing, and reinforcements experimentally, aided by his master mason Edward Strong, whose team built several small-scale masonry models of quarter- and eighth-parts of the dome in 1690–95,[26] and by his ironsmiths Jean Tijou and Thomas Robinson, whose iron chains and ties were crucial for the containment of outward thrusts in the upper stages of the dome.[27]

Near the beginning of the first phase of the design, Strong's team built two masonry models of quarter-parts of the dome (completed in February and May 1691), and Hawksmoor drew at least seven variant designs for a twenty-four-sided dome, one of which he worked up directly from the study design (figure 87, catalogue 46).[28] This sketch develops the concept of a line of thrust from the middle dome to the outer face of the lower drum by ruling these lines in pencil and using them as guides for the outer profile of the middle dome. The parabola becomes flat-sided rather than curved, prefiguring the hidden cone of the completed structure. Wren introduced an upper tier of windows in the peristyle to improve light to the soffit of the inner dome, and, in a more carefully finished design again drawn by Hawksmoor (figure 151), he regularized these windows in twenty-four equal bays and added slots for iron reinforcement in the middle dome and inner wall.

The next group of designs followed a pause of two or three years while Wren gave priority to the construction of the Morning Prayer Chapel at the west end and to designs for the

Geometry and Structure in the Dome of St Paul's Cathedral

fittings of the Choir.²⁹ In drawings datable to *c*.1693–95 Wren narrowed the wall section of the lower drum to about fifteen feet (from about eighteen feet in the earlier group), and infilled every fourth bay as a pier mass, one above each crossing pier, to aid the buttressing of the inner dome (figures 152, 92, catalogue 49). This revision brought the peristyle close to its final form but the drum was low, rising barely above the roofline, the lantern was small, and the upper part of the dome was unresolved, lacking the conical middle dome of the earlier studies. Important evidence for dating this second stage is in Hawksmoor's impressionistic grey-wash sketch in figure 152. This is the most advanced of all his studies for the entire dome before the start of work, and its technique of loosely applied wash over pencil in the upper part of the drawing is characteristic of his hand in the mid-1690s.³⁰ The design of the peristyle in this sketch is close to that in the eighth-part plan of the dome in figure 92. The use of coloured shading on this plan to distinguish the levels of construction and possibly the stone types to be used (the yellow of the peristyle probably signifying Portland stone), suggests that it is a preparatory drawing for Strong's 'large Modell of ⅛ part of the Great Dome', completed in June 1694.³¹ As this was the last new model constructed by Strong, and it included carved work, it must have provided the basis for the executed scheme up to the peristyle.

The fundamental change to Wren's concept for the dome was his decision to slope the wall of his inner drum inwards by one foot in twelve, so that it followed more closely the line

151. Nicholas Hawksmoor, Study for a 24-sided dome, developed from catalogue 46, *c*.1690–91. London Metropolitan Archives, SP 164 (Downes 97)

152. Nicholas Hawksmoor, Sketch study for a 32-sided dome, *c*.1693–5. London Metropolitan Archives, SP 162 (Downes 100)

153. Jan Kip, authorised engraving of the North elevation, 1701. London Metropolitan Archives

of thrust from the triple-dome structure to the crossing arches below.[32] This decision can be connected with a payment in December 1695 for altering and adding to 'the Modell of the Legg of the Dome' (presumably the model of June 1694),[33] a month before Edward Strong's men began preparations for 'Setting out the Foundation of the Dome' above the crossing arches, after which masonry construction at the base of the drum was put in hand.[34] Sloping the inner wall allowed Wren to reduce its thickness to 3 feet 6 inches (from 5 feet in figures 152, 92), increase its height to nearly 80 feet up to the springing of the inner dome, and narrow the diameter of the inner dome, thereby lessening the buttressing requirement of the peristyle and lower drum. He set the sloping inner wall as far back as possible above the soffit bands, and, to reinforce the base of this narrower drum, introduced a honeycomb structure of arches and inverted arches behind the entablature to link the inner and outer drums (figure 150). This hidden arched structure is almost impossible to illustrate in a drawing, and was probably never drawn out in full but simply composed in model form. By connecting the inner and outer walls of the drum at their

154. Exterior of St Paul's Cathedral, from the south

base it helped create what was described in 1930 as 'the great and original box girder which Sir Christopher Wren designed and intended to act as a single unit'.[35]

The executed scheme up to the base of the peristyle was drawn by William Dickinson in a two-part section of the dome and crossing (figure 88, catalogue 47). Dickinson joined the St Paul's office in 1696 and produced this remarkable drawing in two stages over the next four to six years, presumably to assist Wren in judging the visual and structural relationships of the dome to the crossing space, and of the lantern to the outer dome.[36] In the first stage, Dickinson drew alternatives on both sides of the sheet in tentative pen outlines; this dates from before construction reached the base of the peristyle in mid-1700, by which time Wren had slightly enlarged the radial arches of the peristyle to the form shown in Hawksmoor's large-scale working drawing (figure 93, catalogue 50). The profile of the dome and peristyle in the right-hand side of Dickinson's drawing is close to that shown in the authorized engraving of the north elevation by Jan Kip in 1701 (figure 153).[37] What is striking about the right-hand section is the

Compass and Rule

155. Comparative half-sections of Wren's British Museum study sketch (figure 86) and Poley's section of dome as built (figure 138)

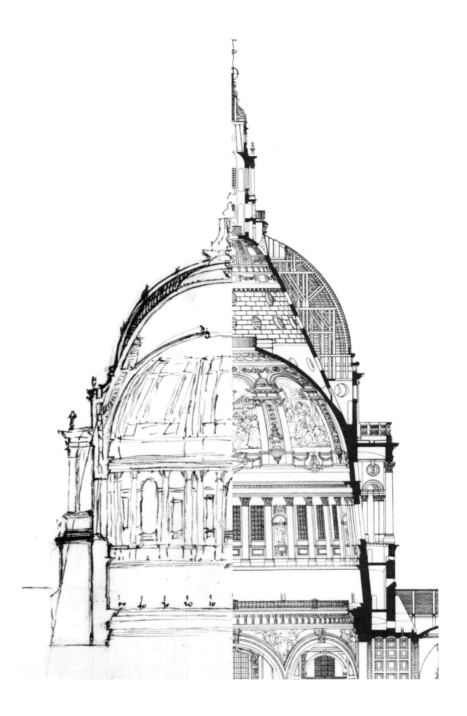

provisional nature of the inner cone. It is straight-sided and rests on a horizontal sill, without any meaningful support directly below. There is no slot for an iron chain at its base, and support for the lantern is shared between the inner and outer domes. This suggests that, when work began at the base of the drum in 1696, the entire superstructure – cone, outer dome, and lantern – was to be in timber.

The second stage was drawn by Dickinson in pink line and shading on the left-hand section. In a radical change to the upper part of the dome, Wren introduced a brick cone that rises from the back of the inner drum to support a much taller lantern of solid, masonry construction that rises above a timber-framed outer dome. Lowering the inner dome blocked the upper tier of windows in the peristyle – windows that are still shown open in Kip's engraving and in Hawksmoor's contemporary working drawing – but created a more satisfactory visual relationship between the inner dome and the crossing space.[38] Begun in 1705, the base of the inner dome serves as a common springing point for a flat-sided cone, eighteen inches thick, reinforced with stone bands and iron girdles at several levels. The cone carries a tall, Portland-stone lantern that appears to rest on the lead-clad outer dome (figure 154).

The revision in Dickinson's drawing can be dated within a year or two of the accession of Queen Anne in March 1702; for, in the summer of that year, parliament granted a generous financial provision for the completion of the cathedral, and this boost in funding – to at least three times the level of a greatly reduced settlement imposed under William III in 1697 – prompted Wren to revise the design of the belfry and lantern stages of his western towers and the uppermost parts of his dome (compare figures 153, 154).[39] In terms of internal structure and external profile, the revision anticipates, in all its essentials, the dome that was completed in 1710.[40]

By setting a large stone lantern of about 700 tons on a masonry-built inner structure, Wren brought a much greater weight down on the fronts of the crossing arches than he had envisaged in 1696. This change in the design at a very late stage had long-term consequences for the stability of the dome. Over the next two hundred years, the disparity in loading between the inner and outer drums caused dislocation between the two shells of the structure, exacerbated by movements in the subsoil. In the early 1920s, fears of a massive structural failure led to a programme of steel and cement reinforcement within the crossing piers and around the base of the drum, completed in 1930.[41]

Comparing the left half of Wren's study-design with Arthur Poley's section of the completed dome (figure 155) it is evident that by shifting the parabolic line of thrust inwards to the fronts of the crossing piers Wren transformed the design of his dome. He was able to raise the lower drum much higher above the roofline than in all the studies drawn by Hawksmoor in the early 1690s, enlarge his outer dome and lantern, and return the peristyle to the classical type of circular temple colonnade that he had first emulated in his design for a dome over the crossing of the medieval cathedral in 1666 (figure 81, catalogue 43).

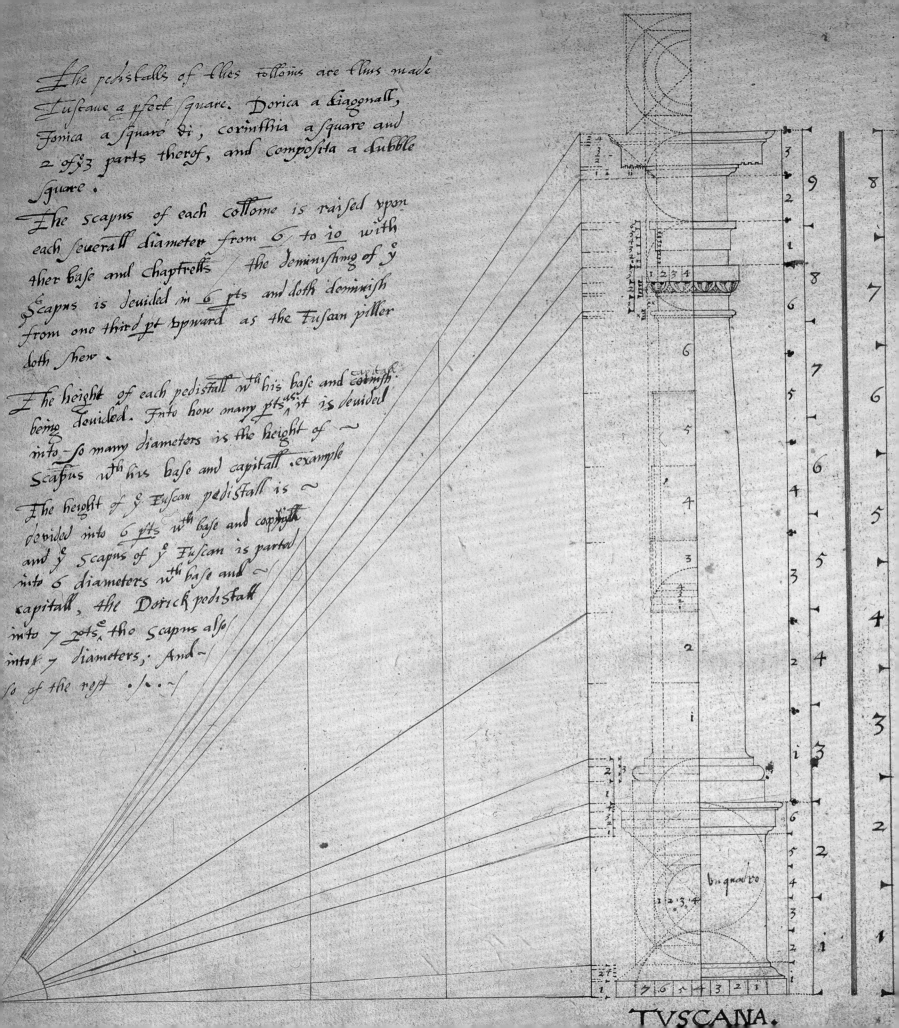

The pedistalls of thes collums are thus made
Tuscane a 1 foot square. Dorica a diagonall,
Jonica a square & ½, corinthia a square and
2 of 3 parts therof, and Composita a dubble
square.

The scapus of each collome is raised vpon
each seuerall diameter from 6 to 10 with
ther base and chaptrells, the deminishing of y
Scapus is deuided in 6 pts and doth deminish
from one third pt vpward as the Tuscan piller
doth shew.

The height of each pedistall wth his base and capitall
being deuided. Into how many pts as it is deuided
into so many diameters is the height of
Scapus wth his base and capitall. example
The height of y Tuscan pedistall is
deuided into 6 pts wth base and copinth
and y Scapus of y Tuscan is parted
into 6 diameters wth base and
capitall, the Dorick pedistall
into 7 pts, tho Scapus also
into 7 diameters. And
so of the rest.

TVSCANA.

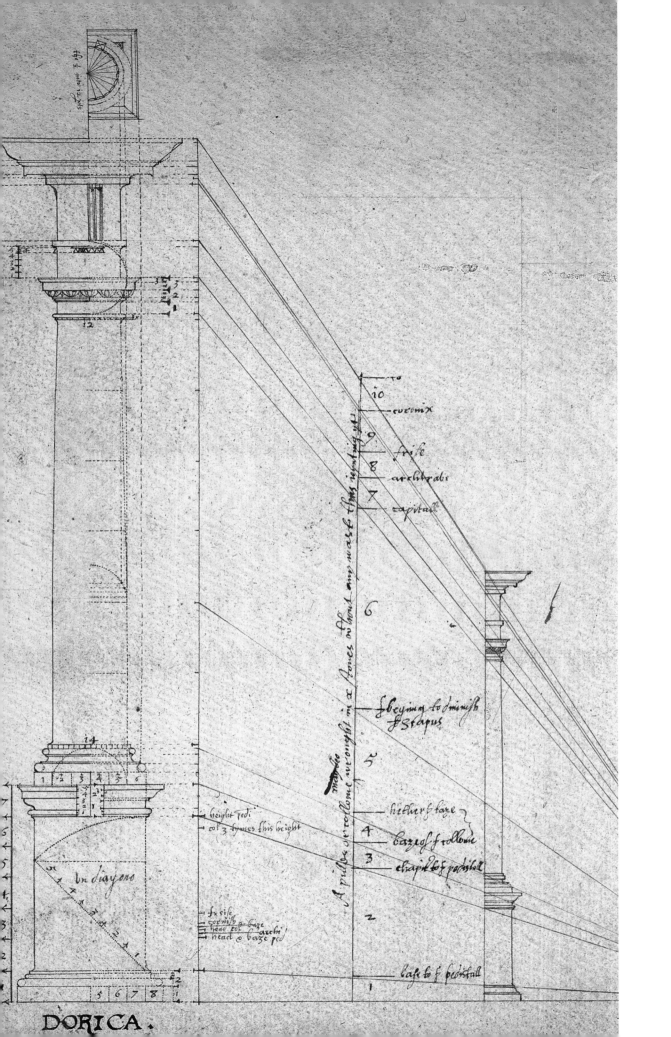

CATALOGUE CHECKLIST

Catalogue Checklist

1. INCISED WINDOW SKETCH FROM THE CHAPEL OF THE HOSPITAL OF ST JOHN THE EVANGELIST

   Late thirteenth century
   University of Cambridge Museum of Archaeology and Anthropology,
   ac. no. Z 15088
   Figure 5

The drawing was discovered in 1869, during the demolition of the old chapel of St John's College. That building had been erected in the late thirteenth century for the use of the Hospital of St John the Evangelist, which then occupied the site. The chapel was remodelled in the years following the foundation of the college in 1511. Most of the original Early Decorated windows were replaced at that time by windows in the Perpendicular style. The old chapel was described and documented by Babington (1874), while Willis and Clark (1988 [1886], ii: 280–308 have traced its subsequent architectural history up to its destruction in the nineteenth century. The chapel's foundations can still be seen set into the grass of St John's First Court. The drawing was mentioned briefly by Coulton (1928, 178), but was not given any extensive analysis or treatment until Biddle (1961), who used it to reconstruct the appearance of the chapel's original east window. The same article also provides a useful list of medieval architectural drawings known at that time. The block has since been reproduced several times.

2. LARGE COMPASS

   Eighteenth century
   Science Museum, London, 1982-907
   Figure 7

Although later in date, this eighteenth-century iron compass is part of the same tradition of manufacture and use practised by medieval master masons. Shelby (1964, 394, n. 37) cites the building accounts for Scarborough Castle of 1429/30, which record 'Also in the trasynghous ii peir of compasses of iren one of the lengthe of a yerd another lesse'. The large size enabled it to be used for drawing and setting out at full scale. Shelby (1965, fig. 2) reproduces a thirteenth-century miniature of a large compass similar in form to this example, whose radius is 1 metre. It has a wing arc of radius 31 cm and the mobile leg can be fixed in place through a screw hole. The hinge is of sandwich construction giving greater friction than simply two blades placed together. It has been marked 'SW' and 'GD' by different owners.

3. BUILDER'S LEVEL

   Nineteenth century
   Museum of the History of Science, Oxford, inv. no. 86404
   Figure 8

Like compasses, levels retained their medieval forms for many centuries. Although spirit levels were introduced in the seventeenth century, the simplicity of the plumb line continued to recommend it for use on building sites. Shelby (1961, fig. 2) reproduces a fifteenth-century miniature with a level almost identical in form to this large nineteenth-century example. Whereas ancient Egyptian and Roman levels were A-shaped and thus assessed the horizontal in relation to only two points, the long horizontal board of medieval examples provided a more reliable check along a course of stonework.

4. William Worcestre (and Benedict Crosse?)
   DRAWING OF THE SOUTH PORTAL JAMB OF ST STEPHEN'S CHURCH, BRISTOL

   August 1480
   Corpus Christi College, Cambridge, Parker Library, MS 210, f. 129
   Figure 9

Worcestre's *Itineraries* were first published in an edition by Nasmith (1778). The part of the manuscript containing the description of Bristol was re-edited by Dallaway (1834). Both of these works contain Worcestre's description of the south portal of St Stephen's, along with his extended list of moulding terms, but neither reproduced the sketch. John Harvey included it for the first time (in his edition, Worcestre (1969)) in the form of a re-drawn diagram, later re-used by Frances Neale in Worcestre (2000). Harvey was also the first to suggest that it may have been drawn by Crosse himself. Pacey (2007, 118–21) reproduced the drawing as a photograph and also provided a measured elevation and section of the portal jamb to compare it. This page has never been exhibited before.

5. PLAN OF THE COURT AND KITCHEN OF WINCHESTER COLLEGE

   c.1394
   Winchester Muniments, 22820
   (inner binding of hall book)
   Figure 11

The plan was published for the first time by John Harvey in Skelton and Harvey (1986, 141–46). The author gives a full account of the drawing, explaining the circumstances of its survival and the odd divergence from the actual buildings that it represents. The fragment is also set into the larger context and literature of medieval architectural drawing on the continent and in England. The inscription at the bottom reads 'locus noue coquina contineret xxx pedes longitudine et in latitudine xx pede'. Emphasizing the lack of scale, Delano-Smith and Kain (1999, 28–29), mention it as a type of medieval map that would have required verbal explanation. The drawing has never been exhibited before.

Catalogue Checklist

6. PLAN AND ELEVATION OF LONDON PLOT AND 'TENEMENTS'

    'Small Register' of deeds, Bridge House Estates Muniments, f. 9r
    c. 1475
    London Metropolitan Archives, CLA/007/EM/04/003
    Figure 15

Published in Harvey (1952), with transcriptions for all four plots. The inscription written sideways up the centre of the page reads: 'This is a patron of ij tenements ate depfordstronde lete to Richard Gilmyn from midsomer a° xvo Regis E iiij^{ti} [a° 1475 added in later hand] vnto the ende of iii^{xx} jxix yeres [expired a°. 1574 added in later hand].' The elevation shows a pair of timber-framed houses with a shared roof, their gables given in perspective view. Harvey also provides a brief analysis of the technical terms used in the inscriptions in the light of other period documents.

7. attr. to William Vertue
    PROPOSAL FOR A TOMB FOR HENRY VI

    1504–09
    British Library, Department of Manuscripts, Cotton, Augustus II.i
    Figure 16

The drawing is lettered at the top in a later hand: 'The Monument intended for Kinge Henry the sixte'. It has been known since at least the late eighteenth century, when it was reproduced in Gough (1786–96, iii: 231–36). The engraving was done by Basire from a drawn copy by J. Schnebbelie. The dating of the design has proved elusive, although most authors have agreed that it relates to the proposed canonization of the late king under Henry VII and the latter's wish – never realized – to translate his uncle's remains from Windsor to the new Lady Chapel at Westminster. Gough mentions both of these circumstances, while also noting the possibility that the design had been made for Henry VIII forty years later. Perkins (1938–52, ii: 149–60) also saw it as a proposal for Henry VII's chapel, adding that the canopied superstructure was intended to enclose a chantry chapel, with the tomb-chest and effigies lying below. John Harvey (1953, and in Harvey and Oswald, 1987, 307–10) dated the drawing to c. 1500 and suggested an attribution to Robert Vertue or, more probably, to his brother William, on the basis of stylistic similarity with the latter's Lupton chantry at Eton College. Colvin (1963–82, iii: 211) mentions it briefly, noting that it is the only surviving graphic record of the chapel or its furnishings. Lindley (2003, 259–93) provided an extended analysis of the funereal iconography and noted the tomb's many stylistic similarities with the Westminster chapel itself, adding further support to the attribution to one of the Vertues. Christopher Wilson, however, in Marks and Williamson (2003, 89, 166–67) has argued that Henry VII had intended to build a shrine in Westminster, not a tomb, and that the proposal must therefore have been intended for Henry VI's resting place in Windsor. He also attributes the drawing to Vertue, but dates it to c. 1515.

Few authors have discussed the representational conventions used in the drawing, in particular the use of parallel perspective. The device was presumably intended to make the rectangular volume of the tomb legible in a way that a pure elevation could not. It is notable, however, that the diagonal lines of the short side of the monument do not converge toward a vanishing point, thereby retaining the same length with respect to the front. In principle, the drawing could still allow a builder to derive the width of the short side from these 'receding' orthogonals. It is another example of a drawing convention derived from practical necessity and closely resembles the technique of 'military' or 'isometric' perspective expressly developed by sixteenth-century fortification engineers to avoid pictorial distortion. On this technique, see Camerota (2004).

8. attr. to William Vertue
    DESIGN FOR THE CHANTRY CHAPEL OF BISHOP RICHARD FOX, WINCHESTER CATHEDRAL

    1513–18
    Royal Institute of British Architects, Smythson SA 51/2
    Figure 17

The drawing was published by Girouard (1956), who identified it as a proposal for Bishop Fox's chantry, noting the proportional differences with the executed version. It has generally been attributed to William Vertue, following John Harvey in Harvey and Oswald (1987, 307–10), on the basis that Fox had employed the same mason to build Corpus Christi College around the same time. Documentary evidence for the dating of the chantry to the years between 1513 and 1518 was discovered by Smith (1988). The sculpture was destroyed in the Reformation. Phillip Lindley (1988) has discussed the programme and the attribution to Vertue. P. D. A. Harvey (1993b, 96, 98), noting the consistent scale, mentions it in relation to the appearance of scale maps and plans. Christopher Wilson in Marks and Williamson (2003, 244–45), has interpreted it as a post-Reformation topographical record, although this view had been considered and rejected by Girouard. Pacey (2007, 168–69) has discussed the perspective effects in the sculpture niches below and on either side of the windows. This is, as he points out, a common convention of the time, intended to make projecting and receding elements legible in the absence of a plan. The drawing was exhibited at the RIBA in 1961 and 1972 and at the Drawing Center in New York in 1983. For the latter, see Harris, Lever and Richardson (1983).

9. PLAN AND ELEVATION OF A CANOPIED NICHE OR PEDESTAL

    unknown date
    Royal Institute of British Architects, Smythson IV/2
    Figure 18

The elevation was published in Girouard (1956) and again with the rest of the Smythson collection in Girouard (1962). It was exhibited at the RIBA in 1984. It has not been possible to determine whether the pedestal and canopy were intended for a specific monument, nor how they would have featured on one. Lever and Richardson (1984, 39, quoting Christopher Wilson), have suggested that it may have belonged to a pair of canopied pedestals flanking an altar, either set into a wall or – as seems more likely from the plan – free-standing.

10. Richard Lee [?]
    VIEW OF THE TOWN AND HARBOUR OF CALAIS

    c. 1541
    British Library, Department of Manuscripts, Cotton, Augustus I.ii.70
    Figures 22, 23

The drawing might be plausibly dated to 1541, after the completion of the Beauchamp bulwark, shown at the north-eastern corner of the town wall. One of the two new towers of the Rysbank fort, at the tip of the spit between the harbour and the ocean, is also shown, but not in the form ultimately carried out. A later drawing of the site, also presumably by Lee (BL, Cotton, Augustus I.ii.57), shows the executed arrangement of the fort more clearly. The additions were begun in February 1541 and completed a year later.

As Barber (1992, 49, n. 60) has observed, these two drawing are very close in style to others in the Cotton collection, including a famous view of Dover Harbour (Augustus

I.i.22–23) and a fragmentary view of the English assault on Edinburgh in 1544 (Augustus I.ii.56). The group can be ascribed to Lee, who was connected to all three sites and who was well known to contemporaries as a draughtsman and cartographer. There is also a rough map of Orwell Haven (Augustus I.i.56) endorsed in his name. A detail of the exhibited Calais drawing was published in Colvin (1963–82, i: 444) as evidence for the form of the Rysbank tower, begun in 1384. The drawing is misidentified there as Augustus I.ii.71.

11. attr. to John Rogers
    PLAN FOR FORTIFICATIONS AT HULL

    1541
    British Library, Department of Manuscripts, Cotton, Augustus I, Supp. 4
    Figures 26, 27

The plan, partially reproduced in Shelby (1967, pl. 8), is one of four surviving drawings of the same project, all in the Cotton collection. The others are Augustus I.ii.49; Augustus I, Supp. 3; and Augustus I, Supp. 20. Unlike the exhibited plan, the others show the long wall preceded by a moat, which we know from later evidence was, in fact, executed. There is no stated scale on the drawing, although dimensions are given for 'The lenght betwene the Bulwarke next the Humber and the Castell' (1066 feet), 'The whole length of the woorke' (2894 feet), and the breadth of the haven at high water (246 feet). The north bulwark is pasted over. The plan is reproduced in P. D. A. Harvey (1993b, 40).

12. SURVEY PLAN OF PORTSMOUTH

    1545
    British Library, Department of Manuscripts, Cotton, Augustus I.i.8
    Figure 28

P. D. A. Harvey (1981) was the first to call attention to the unique cartographic status of this plan. Biddle and Summerson (1982b) situated it in the architectural history of the town and also discussed the significance of the pencilled additions for the history of English fortification. It was reproduced again in Harvey (1993b, 72). A dull olive wash has been used to mark earthen ramparts and buttressing. The draughtsman is unknown.

13. John Rogers
    PROPOSAL FOR REMODELLING THE FIRST FLOOR OF HULL MANOR

    June 1542 or 1543
    British Library, Department of Manuscripts, Cotton, Augustus I.i.84
    Figure 31

Henry VIII acquired the manor as a royal residence and keep in 1539, and Rogers was commissioned to alter it for the king's use soon after. Shelby (1967, 34–46) has reconstructed the sequence of surviving drawings. The others are Cotton Augustus I, Supp. 1 ('Plat A'); Augustus I.ii.11 ('Plat B'); and Augustus I.ii.13 ('Plat D'). He labelled this one as 'Plat C', arguing that it represented a second proposal for remodelling the king's and queen's apartments and was probably never executed. It is endorsed: 'A new plat made by the same Rogers, of the king his hyghnis mannor of Hulle, the xxvth of June.' Although not specifically stated, the scale is one inch to sixteen feet. P. D. A. Harvey (1993b, 98, 100), was the first to point out the early use of scale for a civic building.

14. attr. to John Rogers
    VIEW OF HULL MANOR

    June 1542 or 1543
    British Library, Department of Manuscripts, Cotton, Augustus I.ii.13.
    Figure 32

Labelled 'Plat D' of Hull Manor by Shelby (1967, 34–46), this drawing shows a proposal for modifications to the house, involving the addition of bay windows, fireplaces and external stairs. Of the four surviving plans for the commission, it is closest in conception to Shelby's 'Plat B', although none conforms in all respects. The use of parallel perspective suggests that it was intended as a presentation drawing, presumably for the king himself. This example of the technique provides an interesting comparison with that in the 'Tomb of Henry VI' (catalogue 8), another mason's drawing. The Hull Manor view has also been reproduced in John Harvey (1949, 91) and John Harris (1979, 16, n. 1).

15. Leonard Digges
    A Boke Named Tectonicon
    (London: Thomas Gemini)

    1556
    Bodleian Library, Oxford, Antiq.d.E.1556.2
    Figure 33

The titlepage has a central woodcut providing a diagram of the cross staff in action, with two figures taking the height of a tower.

16. Leonard Digges
    A Boke Named Tectonicon
    (London: T. Orwin)

    1592
    Bodleian Library, Oxford, Tanner 298(2) (annotated copy)
    Figure 34

17. Humfrey Cole
    SURVEYOR'S FOLDING RULE

    1575
    Museum of the History of Science, Oxford, inv. no. 49631
    Figures 38, 39

This is one of four surviving instruments of similar design by the leading English instrument maker of the sixteenth century (see Ackermann (1998, 71–77) and G. L'E. Turner (2000, 139–43). Developed from a simpler craft instrument, the wooden 'carpenter's rule', Humfrey Cole's device is a multipurpose instrument that could be used for surveying, map work, and specialized calculation. It incorporates an inch ruler and scales of equal parts, as well as scales for reckoning areas and volumes. These latter board and timber scales are supplemented by tables whose values are distributed along the first foot of the inch scale; they allow the user to calculate quantities beyond the extent of the scales. When equipped with sights, it could be used to measure horizontal angles and, with the further addition of a plumb bob, vertical angles could be taken with its quadrant and shadow square. Produced in expensive brass and neatly engraved, it demonstrates both the engagement of mathematical practitioners with practical tasks and also their distance from the humble instruments of craft practice (see Johnston (1994a)).

Catalogue Checklist

18. Thomas Gemini
    ASTROLABE FOR QUEEN ELIZABETH

    1559
    Museum of the History of Science, Oxford, inv. no. 42223
    Figure 40

The suspension attachment at the top of an astrolabe is known as the throne and this example carries a regal inscription: 'Elizabeth Dei Gratia Angliæ Franciæ & Hiberniæ Regina'. The reverse is engraved with Elizabeth's coat of arms flanked by the initials 'E R'. On the basis of household accounts, G. L'E. Turner (2000, 17–18) has plausibly argued that this gilt-brass astrolabe was commissioned by Robert Dudley, Earl of Leicester for presentation to Queen Elizabeth. Turner (105–9) fully describes the instrument which consists of a particular astrolabe on the front and a Gemma-Frisius-style universal 'astrolabum catholicum' on the back. The maker's signature, 'Thomas II 1559', is below the quadratum nauticum on the mater, and there is a single surviving plate with a tablet of horizons on one side and a zodiac-calendar, scale of unequal hours, and shadow square on the reverse. The elaborate strapwork rete for twenty-three stars, as well as the italic engraving, recall the style of the Louvain workshops of Gerard Mercator and Gualterus Arsenius and strongly suggests that the Flemish immigrant Gemini learned his trade there.

The instrument also carries evidence of its later history. An inscription on the plate records that the astrolabe was given to the University of Oxford in 1659 by Nicholas Greaves. Intended for the use of the Savilian professors, it was presented in memory of two Savilian professors of astronomy, Thomas Bainbridge and John Greaves. Presumably it came to Nicholas Greaves from his elder brother John, for whom see Francis Maddison's entry in the *Oxford DNB*. It may therefore have been used in John Greaves's expedition to the Levant of 1637–40. Birch (1737, i: ix) reports that Greaves embarked from Constantinople for Egypt but 'being oblig'd to put in at Rhodes, he went ashore, and taking with him a brass astrolabe of Gemma Frisius, because he durst not make use of any larger instrument, for fear of giving suspicion to the Turks, he found the elevation of the pole there to be 37 degrees and 50 minutes'.

19. WOODEN CARPENTER'S RULE

    1648
    Whipple Museum, Cambridge, Wh.0446
    Figures 41, 42

Boxwood one-foot rule with lead ends. The rule is briefly described in Johnston (2006, 253). One face carries inch scales along each edge, while the reverse has a highly unusual layout. A scale of board measure for reckoning areas runs along one edge and continues on the other, from 7 to 36; this would normally be arranged in a continuous line on a two-foot rule. At one end of this face is a so-called table of board undermeasure, which supplements the scale with values for narrow boards. At the other end is an equivalent table of timber undermeasure, despite the absence of a line of timber measure on the instrument. Scales and tables of this kind are discussed by Knight (1988). The curious design of this example may indicate that the maker was copying by rote, without full comprehension. The style of the decoration certainly appears naïve and quite distinct from the conventions of the London community of mathematical instrument makers, suggesting that the origins of this instrument may lie closer to more humble forms of craft practice. Between the two lines of the board scale is the date '1648'.

20. WOODEN CARPENTER'S RULE

    1659
    Science Museum, London, 1954-292
    Figure 43

This two-foot carpenter's joint rule, in pearwood with a brass hinge, is signed 'Richard Assheton 1659', perhaps for the owner rather than the maker. It is the earliest known example of what evidently became a standard pattern. One side carries lines of timber and board measure, with accompanying tables of undermeasure. The outer edge has a double logarithmic line of numbers; first published by Edmund Gunter in the 1620s, the logarithmic line was used with dividers for calculation. The other face includes inches as well as four scales particularly adapted for carpenters and quantity surveyors. Working with dividers, these scales of 'Circumference', 'Diameter', 'Square Equall' and 'Square Within' provide interrelated dimensions. For example, having measured the circumference of a log with string, its diameter and area ('square equal') are immediately found. The 'square within' gives an indication of the area of usable timber left once the log has been squared off.

These four scales were first mentioned in print in John Brown (1661, 107–8), at the end of a chapter titled 'The use of certain lines for the mensuration of superficial and solid bodies, usually inserted on Joynt-Rules for the use of Work-men, of several sorts and kindes'. The implication that they were already known is borne out by this example. The evidence of catalogue 21 suggests that the design may have been created in the later 1650s.

21. BRASS FOLDING RULE

    1655
    National Museums Scotland, Edinburgh, T.1978.92
    Figure 44

Graduated three-foot brass rule with two hinges, so as to fold to one foot. The rule is engraved '1655 Robert Trollap of yorke free Mason'. Trollap was a successful provincial builder (Colvin (1995a, 989–90)) and was surely the owner rather than the maker: the materials, construction and engraving style are all characteristic of professional London mathematical instrument-making of the period. Trollap's ownership suggests the appeal of metropolitan manufacture and modern practical mathematics beyond the capital.

Apart from the greater length and the double-folding format this is similar in its scales to the slightly later Assheton rule (catalogue 20). One side has scales of board and timber measure (5–100 and 7–40, respectively), with a table of timber undermeasure. The other carries inches, a double logarithmic line, and three unlabelled scales. These are for circumference, diameter, and square equal. There is a vacant space where a fourth scale for 'square within' could have been inserted; its absence suggests that the four-scale pattern of the Assheton rule had not yet been devised.

The rule was illustrated in a sale catalogue at Christie's South Kensington, 14 December 1978, lot 269. It is discussed by Morrison-Low (1997) and in Morrison-Low (2007, 43–44).

22. John Symonds
    PLAN OF DOVER HARBOUR

    1583/84
    The National Archives, Kew
    MPF 1/122/2
    Figure 45

The plan focuses on the harbour, with only a schematic street outline for the town. It is marked with the four cardinal directions and is signed 'This Plat of Dover harbrowgh Is 20 Rodes To one ynch / P[er] I Symans' under the scale bar. Symonds illustrates the cliff overlooking the harbour and the buildings stretching along the shore. He depicts the various projecting jetties down at the harbour mouth but the major structure is the right-angled sea wall originating at the town, which was being planned and built at this time. The wall enclosed what was known as the 'pent'. This was designed to retain water that could then be released to flush away the shingle that perpetually threatened to choke the harbour mouth. This scouring action was controlled by a sluice in the cross wall, shown here in its own building over the stream of the River Dour. The plan is briefly mentioned by Summerson (1957–58, 212) in his account of Symonds' life. Its place in the history of the harbour works is discussed in Johnston (1994b, 230–31).

23. John Symonds
    PLAN OF CURSITORS' HALL

    before 1579
    The National Archives, Kew, MPA 1/71
    Figures 46, 47

The plan is now in three separate pieces, which were originally attached. The principal sheet is the ground floor, which is inscribed 'This Plat is After 7 fots to the ynch'. Attached above, as a large flap, was the upper floor of the hall and library. Incised in the ground floor are three lines to make a second flap in the principal sheet. Underneath, an extra sheet was attached to represent the beer cellar and the wood and coal cellar.

This plan was first uncovered by Summerson (1957–58, 213–14), who speculated that the commission had originated with the architect's cousin, Richard Symonds, a cursitor in the Court of Chancery. The plan is undated, though it would probably have been made before the death, in 1579, of Sir Nicholas Bacon, founder and builder of the adjacent Cursitors' Inn. It is reproduced and discussed in P. D. A. Harvey (1993b, 101).

24. Robert Smythson
    PLAN AND ELEVATION OF 'A ROUNDE WINDOW STANDINGE IN A ROUNDE WALLE'

    1599
    Royal Institute of British Architects, Smythson II/33
    Figure 48

The drawing has been published in Girouard (1962) and Girouard (1983, 168–69, 171), who describes its function and possible origin. The drawing may be related to the work of the masons John, Abraham, and Martin Ackroyd of Halifax, who later moved to Oxford (see chapter 4, above). There are two additional similar designs for rose windows in the Smythson collection (II/34), but without the masons' notes. The top inscription reads 'A: Draughte: For the Platte of a rounde: window: Standinge in A: Rounde: walle: Anno: 1599:'. The drawing is further discussed in Lever and Richardson (1984, 41), where the scale is given as 7/12 inch to 1 foot.

25. WILL OF JOHN SYMONDS

    dated 1 June 1597
    The National Archives, Kew,
    PROB 10/174
    Figure 49

Summerson (1957–58) discovered, transcribed, and analysed the registered copy of John Symonds's will (The National Archives, P.C.C. Cobham 61). That copy was written into a volume by a clerk for record purposes. This document is the original probate version. The final leaf carries Symonds's signature, which appears frail and awkward compared to the signature on his earlier plans. Two of the witnesses signed with a mark.

26. Bartholomew Newsam
    SET OF DRAWING INSTRUMENTS

    c.1570
    British Museum, Prehistory & Europe,
    MLA 1912,2-8,1
    Figure 50

A spectacular and very rare set of sixteenth-century drawing instruments in gilt brass, accompanied by a highly decorated presentation case. The sides of the case depict allegorical figures of Peace, War, Poverty, and Abundance. The bottom of the case is signed 'Barthelmewe Newsum' (d. 1593), an English clock-keeper and clockmaker to Queen Elizabeth, who seems to have been active from early in her reign (Jagger (1983, 13–15, 309)). The set has been described by F. A. B. Ward (1981, 86) and holds scissors, two knives, compasses, a sharpening hone, a folding rule and square, pens, a pencil-holder, a pricker, a beam-compass, five sets of dividers, square scribers, and steel-pointed scribers. Some spaces are vacant. Hambly (1988, 152) described Newsam as the maker; G. L'E. Turner (2000, 267–68) has challenged this assumption, wondering whether the set was a gift to Newsam.

27. Humfrey Cole
    ALTAZIMUTH THEODOLITE

    1586
    Museum of the History of Science, Oxford, inv. no. 55130
    Figure 52

The altazimuth theodolite was the most sophisticated surveying instrument of the sixteenth century. Humfrey Cole's version was a development of the 'topographicall instrument' published in Thomas Digges's 1571 edition of his father, Leonard's *Pantometria*.

This example is on loan from St John's College, Oxford, which was perhaps the original purchaser. The vertical semicircle and the horizontal azimuth circle are both signed 'H. Cole. 1586'. These two components had become separated at some point in their history and were reunited in the twentieth century. There is a hint of architectural decoration in the four struts that attach each side of the 'Quadratum Geometricum' to the enclosing azimuth circle. The instrument has been discussed and illustrated by Bennett and Johnston (1996, 56–57), Ackermann (1998, 87–88) and G. L'E. Turner (2000, 168–69).

28. PORTRAIT OF RALPH SIMONS OR SYMONS

    Early seventeenth century (?)
    Emmanuel College, Cambridge,
    Portrait 7
    Figure 56

The Emmanuel College Catalogue entry notes that the painting occurs in inventories of the Gallery from 1719 (Emmanuel College Archives: CHA.1.4 f. 109v). The representation of the inscribed frame (not shown in the figure) suggests that the work may have been copied from an Elizabethan original, which had the inscription painted on a wooden frame, but no such prototype is now known. The painting was described by Willis and Clark (1988 [1886], ii: 475) and first published by Airs (1995, 33). A portrait of Simons recorded at Sydney Sussex College, Cambridge, from 1639 to 1746 is now missing.

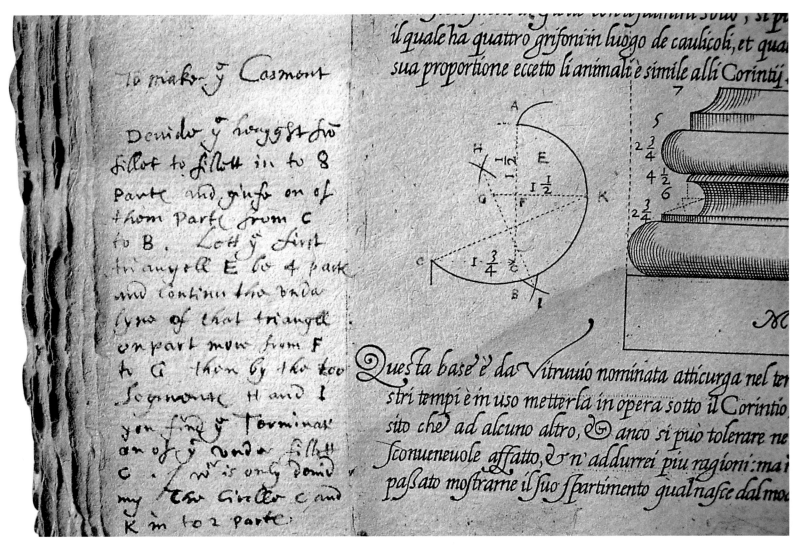

156. Inigo Jones's annotated diagram for the construction of the Attic base (detail of figure 59)

29. Giacomo Barozzi da Vignola
    Regola delli cinque ordini
    (Rome: Andreas Vaccarius)

    1607
    Figure 59

30. Andrea Palladio
    I quattro libri dell'architettura
    (Venice: B. Carampello)

    1601
    Figure 61

31. Buonaiuto Lorini
    Le fortificationi
    (Venice: F. Rampazetto)
    1609
    Figure 62

All with annotations by Inigo Jones
Worcester College, Oxford

George Clarke purchased the bulk of Inigo Jones's library and a group of the architect's drawings from the descendants of John Webb sometime before 1705. The architect's copies of Vignola and Lorini were probably included among this group, but his Palladio came from a different source. As noted on a blank leaf, Clarke bought it from the engraver Michael Burghers on 3 March 1708/9. In 1736, Clarke bequeathed this collection, along with his entire library, to Worcester College, where they remain. The posthumous history of Jones's collection is reviewed in Allsopp (1970), in Harris, Orgel, and Strong (1973, 63–67, 217–18), and in Harris and Tait (1979). On the contents of the library and the current whereabouts of other volumes once in Jones's possession, see Newman (1988) and Anderson (2007). The plan of 'Le Galluce' from Jones's Palladio has been reproduced in Tait (1978) and Tait (1987). The annotations to Lorini are discussed in Harris, Orgel, and Strong (1973, 64) and in Anderson (2007, 77–82).

The scotia of Vignola's Attic base is formed from two circular segments (figure 156). The centre of the first – marked 'F' by Jones – is found by a proportional method based on the division of the moulding into three parts. The second centre – marked 'G' by Jones – is found geometrically, by bisecting CK and extending K to G. Jones's marginal note attempts to simplify this procedure by dividing the moulding into eight parts and ascribing one to the geometrically found length GF. GK, the radius for the lower curve, then allows Jones to find the edge of the fillet, C, where the arc meets the horizontal line one part above the base, B. Working this way, there is no need for stepping off Vignola's 1¾ parts from the edge of the fillet and no need for bisecting CK using points H and I. It appears from the last lines of the note that Jones realized the latter point belatedly. It reads:

> To make ye Casment Deuide ye heayght from fillet to fillett in to 8 Parts and giufe [give] on [one] of them Parts from C to B. Lett ye first triangell E be 4 parts and continu the under lyne of that triangll on [one] part more from F to G then by the too [two] segments H and I you find ye Termination of ye under fillet C. / w[hich] is only deuiding the Circlle C and K in to 2 parts.

The note's italic handwriting can be dated to roughly 1620.

32. 'THE ARCHITECTURAL AND MATHEMATICAL MODEL' OF CLEMENT EDMONDES

    c.1620
    Museum of the History of Science, Oxford, inv. no. 90861
    Figure 63

The first modern interpretation of the model is provided in a short note by Gunther (1923–45, i: 123–24), who observed that the German traveller Uffenbach (1928, 11) had seen it standing in front of one of the windows of the Duke Humphrey Library, probably in the Selden End, during his visit in 1710. Gunther was also the first to realize that the sculpture originally had a pedagogical purpose related to both mathematics and architecture. In support of this view, he cited a 'prospectus' put forward by David Gregory in 1700 for a broad scheme of mathematics teaching at Oxford, in which was proposed a possible lecture on 'the five orders of pillars and pilasters … and other things relating to architecture; as the comparative strength of vaults or arches' (see Jackson (1885, 322)). The latter point is of questionable relevance. Gregory's prospectus was composed long after the model was donated, and it was for a program of extracurricular tutoring, not for the normal lectures of the Savilian professors. Nor, finally, is there any evidence that the architecture lecture was ever delivered. The printed programme of Gregory's classes in the *Mercurius Oxoniensis* (1707) does not mention it, nor does it appear in the version that Gregory had sent for comment to Pepys (1926, ii: 91–94).

Poole (1922) was the first to connect the model with Henry Savile's architectural interests, as reflected in the Bodleian's Tower of the Five Orders. Newman (1988, 154), confirmed this point, with the additional insight that the two works share the same symbolism. Marr (2004) built on these earlier studies, while also casting a much wider net for early seventeenth-century sources. In contrast to previous authors, Marr emphasized the biography of the donor, Clement Edmondes, exploring his lifelong association with the University of Oxford and his own intellectual and professional interests, particularly those related to military history and administration. With respect to the character of the model, the latter point is salient, for it might explain the presence of the rusticated blocks on the faces of the pilasters. Such decoration was, as the author points out, reserved for fortified gates and, in this context, could allude to the donor's belief in the importance of teaching the military arts, especially fortification, within a broad mathematics curriculum. It should be noted, however, that rustication was to be used only with the Tuscan and Doric orders, so its appearance on all five of the model's pilasters therefore represents a real transgression of classical architectural principles. This circumstance and the fact that the number of rusticated segments varies from pilaster to pilaster suggest that they also had a more practical pedagogical purpose, related either to the rules for constructing the orders or to the properties of the geometric solids placed below.

This question brings up another, which Marr also explores: was the model altered after its donation in 1620, with the brass forms, in particular, added later? This does appear to be the case. The entry in the Benefactor's Register does not mention them, specifying only that the model was fashioned from alabaster. Moreover, the brass 'pentakis' dodecahedron at the top of the pillar has been fixed into place in a clumsy way, hanging from a metal rod drilled into the stone. As Marr observes, the workmanship of this addition contrasts greatly with the fine carving in the rest of the sculpture. The ten small pieces of metal jutting up from the base at regular intervals also indicate some intermediate and ill-considered changes to the composition. In the light of these details, it is likely that the two alabaster solids at the base of the pillar are the remaining components of an original set of five, carved from the same material. The brass forms must have been added by 1710; Uffenbach points them out for special attention.

33. PORTRAIT OF HENRY SAVILE

    unknown date
    Museum of the History of Science, Oxford, inv. no. 72703
    Figure 64

This is a copy of the head from the full-length portrait by Marcus Gheeraerts the younger, 1621, in the Bodleian (see Poole (1912–25, i: 32)). The exhibited work was originally lent by the museum's first curator, R. T. Gunther.

34. Euclid
    The Elements of Geometrie,
    trans. Henry Billingsley
    (London: John Daye)

    1570
    Bodleian Library, Oxford, D4. 14. Art
    Figure 66

On this work and the significance of Dee's 'Mathematical Praeface', see Simpkins (1966) and, more recently, Rambaldi (1989) and Mandosio (2003). The cut-out patterns for constructing the polyhedra were most probably inspired by those in Albrecht Dürer's *Underweysung der Messung* (Nuremberg, 1525).

35. John Thorpe
    DRAWING, WITH DETAILS AND NOTES, OF THE CORINTHIAN, IONIC, COMPOSITE, TUSCAN AND DORIC ORDERS
    John Thorpe's Book of Architecture,
    T 13 and T 14

    Sir John Soane's Museum, London, MS vol. 101
    Figures 69, 70

The dating, composition, provenance, and historiography of Thorpe's book are covered in Summerson (1964–66), who also discusses the reconstruction of Blum's orders on T 13 and T 14. The author usefully points out some of Thorpe's departures from Blum's rules, notably in the setting-out diagram for the Ionic volute, which appears to be borrowed from Vignola. Malcolm Airs's recent article in the *Oxford DNB* has summarized subsequent research on Thorpe's career as a designer. The drawings of the orders, however, do not appear to have been discussed since Summerson. The volume has never been exhibited before.

One part of T 14 contrasts with the high level of finish evident in the drawings as a whole, namely the diagram projected onto the diminishing lines to the right of the Doric order. This schema breaks the column into ten parts, each descriptively labelled, for example, 'base to ye pedestal' and 'chapit' to ye pedestall.'

Catalogue Checklist

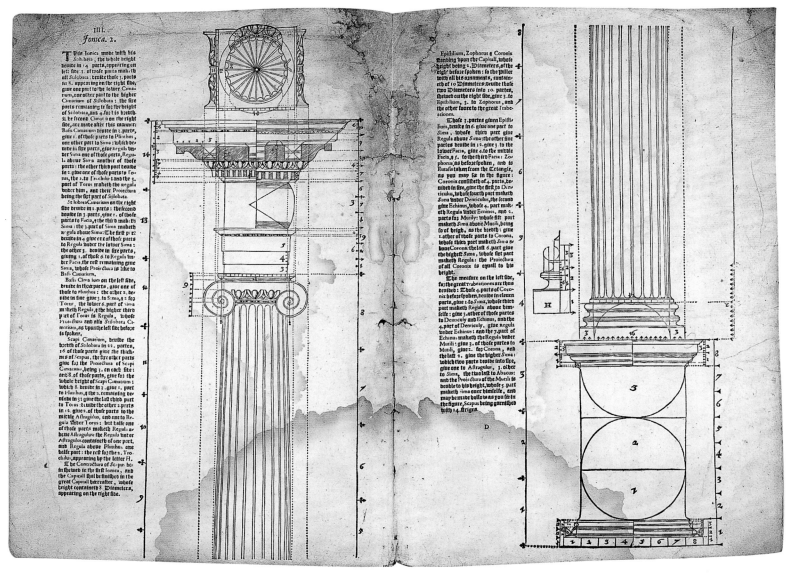

157. 'Ionica 2' from Blum, *The Booke of Five Collumnes* (1601)

Thorpe's purpose was to show the most economical masonry jointing for the order, each part representing a separate block of stone. As he put it, 'A piller or collome may bee wrought in x stones wtout any wast thus jointing yt'. This diagram appears to have been added in a later hand and as an afterthought. Not only does it introduce an element of casualness in what is otherwise a highly finished drawing, it also concerns the execution of the order, as opposed to its design – a perspective that is lacking from Blum's work. If Thorpe had originally planned to use the drawings as the basis of an engraving, this note – which reads as something of a personal observation – may have been added after those plans fell through.

36. Hans Bloome [Blum],
The Booke of Five Collumnes of Architecture
(London: for H. Wounteel, title page)

1601
Worcester College, Oxford
Figure 71

Historians have long believed that this English translation was first published in 1608. An earlier edition of 1601, however, came to light in the 1980s. Its details – known from this sole surviving copy in Worcester College – were published in Eileen Harris (1990). It is an unbound, incomplete version. Apart from the title and dedication pages, the only remaining sheets are those for the Tuscan order and Blum's second version of the Ionic. It is tempting to speculate that its fragmentary condition reflects the wear of continued use (see figure 157). Blum's little book was, after all, a practical manual, and this may be the reason why no other copies of the 1601 edition survive. The title page is inscribed with George Clarke's monogram initials 'G.C.'.

37. Jacques Androuet Du Cerceau,
    Leçons de perspective positive
    (Paris: M. Pattison)

    1576
    (with John Thorpe's MS translation)
    Bodleian Library, Oxford,
    LL 23★ Art. Seld.
    Figure 72

This manuscript was attributed to Thorpe by Höltgen (1990) on the basis of the signature found below the first translated 'lesson'. The autograph matches several examples in the Soane Museum's Thorpe album. On perspective books in England, see Gent (1981) and Anderson (2003).

38. Christopher Wren
    STUDIES FOR A WEATHER CLOCK

    *c.*1663–64
    All Souls College, Oxford, IV.148
    (Geraghty 384)
    Figure 73

Sekler (1956, 32) mentioned this drawing in passing. It has never been exhibited before.

39. Christopher Wren
    DIAGRAM OF THE 1664 COMET

    January 1665
    All Souls College, Oxford, I.3
    (Geraghty 391)
    Figure 75

The inscription on the lower diagram, 'January 1655', should read 'January 1665'. The drawing is based on a theorem that Wren first proposed to John Wallis on 1 January 1665: supposing the comet's motion to be uniform and rectilinear, the problem of finding its path, on the basis of four observations from a given location, is that of finding the straight line cut by four given straight lines, so that its three portions are related to a given ratio. The ratio is that of the time intervals between the observations. See Bennett (1982, 65–66). Wren presented the drawing to the Royal Society on 1 February 1665. According to the minutes in Birch (1756–57, ii: 12, 32), he 'produced some observations of the comet, with a theory.' Hooke, who appears to have been in possession of the drawing, was asked to make a copy and return it to Wren on 19 April. Hooke later published the diagrams in his *Cometa*: see Hooke (1678, 41–43, pl. 5). The drawing has also been reproduced in Bolton and Hendry (1924–43, xii: 24, pl. 47), who published an explanatory letter from the astronomer Andrew Crommelin, noting the discrepancies between Wren's observations and modern data for the comet's path. The drawing was exhibited in 1982: see Downes (1982, 42, 48–49).

40. attr. to Christopher Wren
    WATERCOLOUR SKETCH OF A PIECE OF SMALL INTESTINE

    *c.*1663(?)
    Wellcome Library, London, no. 570945i
    Figure 77

The Wellcome Library acquired the drawing on 5 December 1929 from Oppenheim & Co, a London bookseller. It was first published in William C. Gibson (1970, 336), who described it as showing 'a section of small intestine which has been opened [on the right] to reveal a haemorrhagic ulcer of the type described by Willis in his work on "Putrid Feaver"', that is, in the second book (chapter 10) of his *Diatribae duae medico-philosophicae*, 1659. This characterization is supported by a note in the Wellcome's catalogue record, which reports the opinion of a pathologist at Glasgow University that the drawing 'almost certainly represents typhoid fever of the lower ileum'. Could Wren have been working with Willis at such an early date?

The fact that the drawing had never been attested before 1929 raises some questions of attribution, nor is Wren known to have collaborated with Willis in this particular area. On the other hand, there is little in the drawing itself to provoke outright suspicion. The signature closely matches known examples of Wren's writing from the 1660s and '70s (the Willis endorsement appears to be in the same hand), and a similar skilful use of colour is apparent in his pre-fire designs of 1666. The paper carries half of a fleur-de-lys watermark, which is probably too generic to date the paper reliably. The drawing was exhibited in 1982 and reproduced in the catalogue: Downes (1982, 46, 47).

41. Christopher Wren
    CLASPED HANDS

    unknown date
    All Souls College, Oxford, IV.155
    (Geraghty 389 recto/verso)
    Figures 78, 79

42. Christopher Wren
    PRELIMINARY SCHEME OF THE SHELDONIAN THEATRE

    1663
    All Souls College, Oxford, IV.80
    (Geraghty 1)
    Figure 80

This drawing was recently rediscovered by Geraghty (2002). It has never been exhibited before.

43. Christopher Wren
    PRE-FIRE DESIGN FOR ST PAUL'S, WEST ELEVATION OF THE TRANSEPTS AND DOME WITH SECTION THROUGH THE NAVE

    1666
    All Souls College, Oxford, II.6
    (Geraghty 46)
    Figure 81

This drawing was published in the first volume of the *Wren Society*: Bolton and Hendry (1924–43, i: pl. VII). Sekler (1956, 111–12) was the first to relate the design to precedents in Serlio and to the domes of the Sorbonne and the Jesuit church of St Louis. Inigo Jones's classical masonry cladding, introduced in the 1630s, is visible on the exterior of the transept arms. Note his window surrounds, urns, and finials. Documents related to the pre-fire activities of the Commission for Repairs of Old St Paul's are to be found in Bolton and Hendry (1924–43, i: 13–19, 44–45). For further context, see Soo (1998, 34–55), Jardine (2002, 223–33), and Higgott (2004a, 183–86). For a full bibliography, see Geraghty (2007, 47). The drawing has been exhibited twice before: Downes (1982, 15, 74) and Downes (1991, 28–29).

Catalogue Checklist

44. Christopher Wren
PROJECT FOR THE REPLANNING OF THE CITY OF LONDON

September 1666
All Souls College, Oxford, I.7 (Geraghty 396)
Figure 82

This plan is the second of two versions produced in the immediate aftermath of the Great Fire and presented to the king by 11 September 1666. The plan was first reproduced by Wren (1965 [1750], 268–69), who describes his father as having taken 'an exact Survey of the whole Area and Confines of the Burning, having traced over, with great Trouble and hazard, the great Plain of Ashes and Ruins'. Downes (1982, 63), however, has suggested that he might simply have traced over an existing map. The perimeter of the fire-damaged area is edged in yellow wash. The architectural precedents were first analysed by Sekler (1956, 59–62). For further context, see Reddaway (1951, 40–67) and Jardine (2002, 247–67). For a full bibliography, see Geraghty (2007, 256). The plan has been exhibited in 1982 and 1991, for which see Downes (1991, 29).

45. Christopher Wren
STUDY DESIGN FOR THE DOME OF ST PAUL'S CATHEDRAL, IN TWO HALF-SECTIONS, EACH INCORPORATING THE CURVE OF A CUBIC PARABOLA

c. 1690
British Museum, Department of Prints and Drawings, reg. no. 1881-06-11-203
Figure 86

The drawing was initially published in Dircks (1923, 53), and in 1938 in Bolton and Hendry (1924–43, xv, dedication page), but in both cases without comment. Peach and Allen (1930) were the first to recognize its structural significance, while also relating the shape of the parabola to contemporary interest in the catenary. They also sought to test whether the shape helped to determine the form of the completed work. In an inspired moment, the authors turned a measured section of the cathedral upside-down and hung loops of chain from points along the base. They then printed a photograph of the ensemble right side up. In a similar experiment using an inverted section, Hamilton (1933–34) fitted the suspended chain with a weight to simulate the point-load of the lantern. Both demonstrations were inconclusive. Neither shows any strict relationship between the angle of the suspended chain and the inclined walls of the drum or cone, but the idea was nevertheless suggestive and photographs of both trials have since been reproduced several times.

The provenance of the British Museum drawing is discussed in Croft-Murray and Hilton (1960–63, i: 550), where the lower inscription is attributed to a previous owner, Ralph Thoresby, but without support. Higgott's attribution of the inscription to Hawksmoor is based on extensive experience with the architect's hand. For an analysis, see Higgott (2004b). The drawing was reproduced again in Sladen (2004), where it is discussed in relation to the cathedral's interior decoration. Further structural issues surrounding the dome have been usefully treated in Campbell and Bowles (2004) and Campbell (2007, 138–44).

46. Nicholas Hawksmoor
HALF-SECTION AND ELEVATION OF TWENTY-FOUR-SIDED DOME, ST PAUL'S

c. 1690–91
London Metropolitan Archives, SP 165A (Downes 99)
Figure 87

The drawing was first published in 1926 in Bolton and Hendry (1924–43, iii: pl. XXVII (left)). It is one of seven variant designs drawn by Hawksmoor for a twenty-four-sided dome, succeeding the sixteen-sided model of the 'Revised' design. All of the variants incorporate a conical inner dome, a feature that appears for the first time in this group and that most probably derives from the parabola shown in Wren's British Museum study-sketch (catalogue 45). The current item shows this affiliation most clearly in the buttressing underneath the peristyle and in the pencilled additions to the inner wall of the drum, both of which develop the idea of containing a notional line of thrust. The crossed pencil lines passing through the oculus of the inner dome relate to Hawksmoor's interest in the way that light would fall inside the dome – a feature shown vaguely on Wren's source drawing but elaborated here.

47. William Dickinson
SECTION OF CROSSING AND DOME WITH ALTERNATIVES, ST PAUL'S

c. 1696–*c.* 1702
London Metropolitan Archives, SP 186/1 (Downes 158)
Figure 88

The drawing, first published in 1936 in Bolton and Hendry (1924–43, xiii, p. xv and pl. XIII), was made in two stages. The first stage, which was completed soon after Dickinson joined the St Paul's office in 1696, shows two variant sections of a triple-shelled structure for the dome of the cathedral. The right-hand variant is more clearly visible because that on the left has been obscured by a later modification in pink ink. The elements in the right-hand section – inner, middle, and outer domes – are much taller in proportion than those in the left and in the work as built. As Bolton first observed, the springing of the cone carrying the lantern is placed too high to be contained effectively by the great chain, shown here inserted at the level of the balustrade above the peristyle. In this connection, Higgott, in his contribution in this volume, also notes the unusual construction of the 'cone' on the right-hand side, particularly the thinness of the lines and the beam-like character of its support. Could these features point to a timber superstructure?

The contrast made by the pink additions on the left-hand side tends to reinforce this suggestion. The diagonal section of the cone is brought down past the horizontal sill to rest near the shared springing of the inner dome, where it is further supported by a stepped buttress hidden by the balustrade of the peristyle. In addition, the base of the cone on the left is much closer to the slot intended for the great chain, marked by a small rectangle embedded in the masonry. The pink hatching surrounding that slot is carried upwards to mark out the inner dome and cone as continuous masonry construction (in two double courses of brickwork in the fabric). Finally the lantern has been increased in size and made to rest more squarely on the cone. The revisions are likely to have been made sometime soon after the accession of Queen Anne in March 1702.

As Bolton originally noted, the nave bay adjacent to the left-hand crossing arch is much too narrow. The editors explained this discrepancy by positing the existence of an otherwise unknown plan for the nave but, as Downes (1988, 153), has argued, it is much more likely to be a drafting error. (*This entry was written with Gordon Higgott.*)

48. 'Unidentified' draughtsman, with Nicholas Hawksmoor
STUDY FOR UPPER TRANSEPT END, ST PAUL'S

c. 1685–86
London Metropolitan Archives, SP 132 (Downes 46)
Figure 91

This drawing is the last of a sequence of four studies for the upper transept end (Downes 43, 44–45, 46). It was reproduced in Bolton and

Hendry in 1936 (1924–43, xiii: pl. VII) and discussed by Downes (1988, 25, 33–34, 74–77), who traced the revisions from one variant to the other. Higgott (2004b, 537, 540) observed this draughtsman's hand in more than twenty drawings from Wren's St Paul's office from 1675 to about 1687, although his identity remains uncertain. He argued that the sheet represented the final stage of a collaboration between the 'unidentified' draftsman and the young Nicholas Hawksmoor in the drawings for the upper transept front, dating the sequence to 1685 or 1686. This draughtsman may have been Edward Strong senior (1652–1724), Wren's leading master mason at St Paul's. The hand – characterized by flatly drawn Corinthian capitals, a preference for red-chalk shading on plans, and quirks in the numbering of scale bars – is found in drawings from the beginning of construction in 1675, when Strong himself began work at St Paul's (see below, catalogue 49). Moreover, the west side of the north transept, which corresponds to the right-hand part of the façade shown in the elevation, was one of five areas of construction at the cathedral for which Strong was responsible (see Campbell (2007, pl. 7) for diagrams showing the areas contracted by the master masons at St Paul's). On this particular sheet, Hawksmoor drew the scale bar and inscribed the dimensions while his colleague drew out the final version of the design in ink, presumably as a fair copy for Wren's approval.

Edward Strong senior's possible role as a draughtsman would be analogous to that of Edward Pearce (or Pierce, c.1630–95), the master mason whose hand Higgott (2004b, 537, 541) has also identified among the St Paul's office drawings alongside Hawksmoor's. In one important case, Pearce drew a half-section through the main vault and screen wall in c.1685–86, with Hawksmoor then making a fair copy from Pearce's original (Downes 57, 48). Edward Woodroofe, too, had been brought up in the mason's craft. The presence of these men points to an important transition in the working practice of the office. In the first years of construction, Wren had to rely on masons, not only to execute the work but also to develop designs on paper. Their influence waned, however, as the project progressed. From the mid-1680s Wren began to bring in and train new members of the office – from outside the mason's craft – as 'specialized' draughtsman. The early careers of Thomas Laine (*fl.* 1677–83), William Dickinson (c.1671–1725), and especially Nicholas Hawksmoor mark the beginning of a process of modern professionalization of drawing practice. (*This entry was written with Gordon Higgott.*)

49. 'Unidentified' draughtsman, with Nicholas Hawksmoor
QUARTER-PLAN OF CROSSING WITH VARIANTS FOR PERISTYLE, ST PAUL'S

c.1675–87 (crossing plan) and c.1690–94 (variants for peristyle, by Hawksmoor)
London Metropolitan Archives, SP 11
(Downes 106)
Figure 92

The plan, originally reproduced in Bolton and Hendry (1924–43, ii: pl. IV), was completed in two phases. The base drawing dates to 1675–c.1687 and relates to a larger group of crossing studies produced during the same period (Downes 102–106). Higgott (2004b, 537) attributed all but one of these (Downes 105) to the 'unidentified' draughtsman, on the basis of the scale bar on the left (see catalogue 48, above).

The drawing was completed some years later by Hawksmoor, who turned the sheet through 90 degrees to sketch out two variants for the peristyle on one-eighth plans of the dome. He then worked up the bottom one with a colour wash. This part of the drawing dates to c.1693–94, before the form of the peristyle had been fully determined. The radial buttresses, for example, extend the full 26-foot depth of the supporting arch, rather than roughly 24 feet 6 inches as in the fabric. It is possible that the drawing was preparatory for a masonry model, with the coloured washes used to indicate different stone types. (*This entry was written with Gordon Higgott.*)

50. Nicholas Hawksmoor
LARGE-SCALE SECTION THROUGH DRUM AND PERISTYLE

c.1700
London Metropolitan Archives, SP 113
(Downes 159)
Figure 93

Reproduced in Bolton and Hendry (1924–43, iii: pl.V) and Downes (1988, 154). Most of the pencil annotations and markings by Hawksmoor relate to the dimensions of the masonry courses. Two notes written beneath the base of the peristyle on the left-hand side, level with the base of the pilasters of the inner drum, allow the drawing to be dated to c.1700. One says 'a little of this course sett' and another 'outside levell'. Both refer to the level of construction at the time that the pencil note was made. The notes, however, are additions to the pen drawing, which could date from slightly earlier. The level in question is just above the shaded arch of the void, beneath the chamber that is directly below the floor of the peristyle.

The drawing matches the fabric but for two exceptions: the springing of the inner dome and cone is still at the higher level shown on the right-hand alternative of catalogue 47, so there is a tier of windows above the entablature of the inner drum, instead of the blank panels in the fabric. In addition, there are errors in the alignment and profile of the soffit of the arch at the base of the section. The former is revealing of Wren's working methods. This large-scale construction drawing demonstrates that, when the lower part of the peristyle was being built, in 1700–02, Wren changed his mind about the design of the upper part. The revision is recorded in pink ink on the left-hand side of catalogue 47, where the windows are blocked. Of course, the peristyle in catalogue 47 pre-dates the version in this drawing. It is narrower by some eight inches and lower by about a foot in relation to the capitals of the internal order. The variation, however, does not appear to have mattered, as the only change necessary to the upper part of the peristyle (above the capitals of the inner order, reached by the builders in 1703), was simply to block the openings of the windows while retaining their external frames.

The second inaccuracy is in the treatment of the soffit band of the crossing arch in relation to the sloping wall of the inner drum, where the former is set too far forward with respect to the latter. In the fabric, and in catalogue 47, the rear face of the drum wall is aligned with the rear edge of the inner arch soffit band. The discrepancy does not appear to reflect an otherwise unknown design for a narrower inner drum but is more likely to have been a simple draughting error, resulting from Hawksmoor drawing an entablature that raked back too steeply. This gave him too little space for the soffit band of the arch. Dickinson made a similar error in catalogue 47, where the nave arcade on the left side is too narrow. As in that example, the error was inconsequential, as construction had long superseded that part of the drawing. (*This entry was written with Gordon Higgott.*)

Catalogue Checklist

51. Henry Wynne
    MAGAZINE CASE OF MATHEMATICAL INSTRUMENTS

    Late seventeenth century
    Jesus College, Cambridge
    Figure 99

The contents of this grand magazine case were listed in Gunther (1937, 49–50) and illustrations of the complete set were reproduced in Hambly (1988, 162–65). The case includes a variety of compasses, pens, and rules, as well as a sector, a sundial, and other devices spread over four trays. Although the set was a bequest to Jesus College by Roger North, it is notable that it does not contain any of the personal inventions and adaptations so carefully described in his memoirs: North (2000, 134–38). Aside from a miniature drawing table similar to an example discussed in the autobiography, we seem to have standard products of the trade rather than the results of North's ingenious affection for mathematical recreation.

There is a simple explanation for the lacunae. In discussing 'Ingenuitys' in his brother's biography, Roger recalled that Francis North patronized Henry Wynne, 'a famous instrument maker' and near neighbour in Chancery Lane. Francis

> was a true lover of arts, and as well for the encouragement of that Mr Winn, as for his owne speculative humour, (for he had not time to practise drawing), he caused a case of mathematicall instruments to be made by him yet extant and intire which cost £50 and nothing of the kind can be made by the hands of men more nice, elegant and curious than those are. (Chan (1995, 247–48); for Wynne, see Clifton (1995, 307))

There can be little doubt that the set in Jesus College first belonged to Francis North. Although the case does not capture Roger North's more intense and direct engagement with drawing, architecture, and mathematical practice, it is nonetheless a remarkable and magnificent witness to the gentlemanly interest in instruments.

52. FORTIFICATION SECTOR TO THE DESIGN OF JONAS MOORE

    c.1673
    British Museum, Prehistory & Europe, reg. no. MME 1998,10-2.1
    Figure 100

Jonas Moore was a highly successful mathematician, who began his practical work as a surveyor of the Fens and rose to the post of Surveyor General of the Ordnance and a knighthood; his career is discussed in Willmoth (1993). In addition to his administrative duties, Moore was a patron, supporting the foundation of the Royal Observatory. He also took his military responsibilities seriously, publishing *Modern Fortification* (1673) during the Third Dutch War. That text presented Moore's design of a sector specially adapted for fortification. A. J. Turner (1987, 157–60) introduces the history of the sector, while Bryden (1998) has given a full account of Moore's design. In addition to general-purpose scales, the most distinctive feature of the instrument is the presence of a series of lines for various contemporary systems of fortification, drawing from a range of European authors. One of these proportional schemes, for the design of angle bastion outlines, is attributed to Charles II. This sector is one of several surviving examples in ivory.

53. Thomas Tuttell
    MATHEMATICAL PLAYING CARDS

    1701
    British Museum, Department of Prints and Drawings, reg. no. 1896,0501.925.1-52 (Schreiber E.67)
    Figure 101

One of the more entrepreneurial products of an early modern mathematical instrument maker was Thomas Tuttell's set of mathematical playing cards, issued in 1700. These cards (58 × 92 mm) give a remarkable insight into the wide claims of mathematics at the time and the commercial and cultural territory that its practitioners sought to claim. Five cards have been selected as particularly appropriate to our themes: the King of Clubs (which represents 'Building'), the Queen of Clubs (a drawing table), the Ace of Hearts ('Mathematicks' in general), the Five of Diamonds (an architect), and the Ten of Diamonds (a bricklayer, ironically worth exactly twice as much as the architect). Tuttell's work and the production of specifically mathematical playing cards at this time are discussed in Bryden (1993 and 1997).

54. Samuel Saunders
    PORTABLE CASE OF DRAWING INSTRUMENTS

    1730s
    Collection of Howard Dawes
    Figure 102

The case is shagreen-covered with extremely elaborate applied decoration. One side of the lid carries a badge inscribed 'ARCHITECTURA', the other has 'GEOMETRIA PERSPECTIVA'. The openwork foliage on the main body of the case frames two groups of three *putti*, each group arranged around a globe. The *putti* variously hold mathematical instruments and geometrical solids, and one rests on an architectural capital. The case contains silver and steel instruments, with a silver-mounted ivory sector, a rectangular protractor and plain scale, a pencil holder, dividers, and compasses with a removable dry point, extension attachment, and ink and wheel pen blades. There is also a blade and adjustment tool. The case and instruments are signed by Samuel Saunders, who is noted by Clifton (1995, 244) as active from 1708 to 1743 and described as both a toyman and a mathematical instrument maker.

55. Thomas Heath
    ARCHITECTONIC SECTOR TO THE DESIGN OF THOMAS CARWITHAM

    c.1723
    Museum of the History of Science, Oxford, inv. no. 25362 purchased with a grant from the PRISM Fund.
    Figure 103

Thomas Heath was both the instrument maker and the publisher of Thomas Carwitham's architectonic sector. Only two examples of the instrument are known: a brass version that closely matches the plate in Carwitham (1723), now in a private collection; and this 9" radius example in silver, which differs in size and small details from the published version (described in Tesseract (2005, no. 24)). The instrument's primary purpose was to encapsulate and rescale the proportions of the flutes and fillets of pilasters and columns in both plan and elevation. The most distinctive feature is the small additional hinged leg, which makes it possible to work on two sets of proportional operations at the same time. Although intended for architects and draughtsmen rather than carpenters, the restricted range of tasks targeted by Carwitham's instrument recalls the deliberately limited reach of John Brown's joint rule, as published in Scamozzi (1669). That Brown's earlier sector was taken as a model is also suggested by the way that Carwitham referred to his scales. Echoing Brown's 30 and 40 lines, Carwitham's sectoral scales were referred to by the number of their parts as lines of 29 and 30, and 32 and 48.

56. Thomas Heath
ARCHITECTONIC SLIDING PLATES TO THE DESIGN OF THOMAS CARWITHAM

*c.*1723
Collection of Howard Dawes
Figure 105

When Thomas Carwitham described his architectonic sector in 1723 he also discussed an alternative format for the same sectoral scales, intended for use with smaller dimensions and requiring a special pair of dividers. This instrument is a unique example of Carwitham's architectonic sliding plates. When closed, it resembles a short plain scale, with the main body carrying on one side the signature 'Tho: Heath Fecit' and inches divided into parts, from 1 to 60. The reverse has a diagonal scale and proportional scales labelled 'Decimal part of a / foot Inch & parts' and 'An Inch & parts Divided in 12 E[qual] P[arts]'. These are simply practical additions occupying otherwise vacant space. The sectoral scales appear on each bevelled edge and on thin sectors that are hinged to be pulled out of the main body. With more space available, Heath has been able to provide more informative titles for each pair of scales, such as 'F^r ye Flut^s of a Col^n in Plano'.

57. William Halfpenny
Magnum in Parvo, or, the Marrow of Architecture (London: for John Wilcox and Thomas Heath)

1728
Collection of Howard Dawes
Figures 106, 108

As documented by Eileen Harris (1990), this volume is one of several from the early part of William Halfpenny's career to teach the construction and proportions of the five orders. The particular feature of the treatment here was its introduction of a pair of architectural protracting instruments that embodied the proportions of the orders on engraved scales. Their manufacture was entrusted to Thomas Heath, who figures on the title page as one of the publishers. Both the book and instrument would have been available from Heath's shop 'next the Fountain Tavern in the Strand'. Halfpenny was based close by. His address is given as 'Exeter-Change in the Strand' on the title page of his *Art of Sound Building* (1725), although in that volume he recommends not Heath but a more senior contemporary:

> All Gentlemen that have occasion for such Mathematical Instruments as are used in Architecture, I recommend them to Mr Thomas Wright, (Mathematical Instrument-Maker to his Royal Highness George Prince of Wales) at the Orrery and Globe next the Globe-Tavern in Fleet-Street, in Justice to his great Care in making them very accurately, and for his great Choice of Mathematical Instruments in general. (concluding advertisement)

Halfpenny's switch to Heath was evidently a success. Three years after *Magnum in Parvo* he issued *Perspective Made Easy: or, a new method for practical perspective shewing the use of a new-invented senographical protractor* (1731), to which he added the note

> Since Drawing depends much on good Instruments, and particularly on that described in this Tract, as well as all Mathematical Instruments; I think I do the World service in recommending Mr Thomas Heath, Mathematical-Instrument Maker, at the Hercules and Globe, next door to the Fountain-Tavern, in the Strand; whose Character is well known to the Practitioners of the Mathematical Art. (vi)

58. Thomas Heath
ARCHITECTURAL PROTRACTORS TO THE DESIGN OF WILLIAM HALFPENNY

*c.*1728
Collection of Howard Dawes
Figure 109

These are the only surviving examples of the instruments described in William Halfpenny's *Magnum in Parvo* (1728) and depicted in a folding plate. They were intended to slide within grooves along two edges of a specially adapted drawing board. Both instruments have a pair of surprisingly crude cuts at two corners. These were presumably an afterthought and suggest that the devices were prone to slip in their grooves; the cuts would have been used to give the corners some spring tension and keep them in place while drawing. Both instruments conform closely to the pattern illustrated in Halfpenny's book. The larger is inscribed on the rotating sector 'W^m ★ Halfpenny Invent' and 'Tho. Heath Londini Fecit'. The smaller is double-sided, carrying scales for Tuscan, Doric, and Ionic on one side and Corinthian, Composite, and equal parts on the reverse.

59. Thomas Heath
ARCHITECTONIC SECTOR TO THE DESIGN OF OTTAVIO REVESI BRUTI

*c.*1737
Museum of the History of Science, Oxford, inv. no. 55213
Figure 111

The design was initially published by Bruti (1627) and has been rather fantastically discussed by Carpo (2004) as a precursor of punched card machinery. However, like other sectors, it is used with dividers, which are used to take off dimensions from the sectoral scales on the legs. The separation of the legs is adjusted over the scales on the broad arc or limb.

Malie (1737) is an English translation published by Thomas Heath, who also made the instrument. This brass example is unsigned but can be confidently attributed to Heath; it is the only known example of the design. One side of the arc is primarily for heights: arches and colonnades; doors and niches (with their widths); basements; and entablature. The reverse is for widths: colonnades and arches; basements and entablature; cornices of doors; and scrolls and balusters. Each group of scales (or 'interstices') consists of five lines, marked 'T', 'D', 'I', 'R', 'C' for the five orders and further subdivided and dotted with register points to allow the legs to be set. For example, the group for the widths of colonnades and arches is subdivided into four segments for colonnades with and without pedestals, and arches with and without pedestals. The only exception to the general pattern is the final 'interstice' for scrolls and balusters, whose five lines are marked 'F', 'V', 'S', 'B', [blank] for flutes, volutes, scrolls and their thickness, and balusters or colonettes. The book notes that 'The fifth and last having no divisions marked upon it, is left to the discretion of the ingenious Artist, to improve as he shall think proper, according to his skill in Architecture or Mathematicks'.

Catalogue Checklist

60. John Robertson
    Treatise of Such Mathematical Instruments, as are Usually Put into a Portable Case
    (London: T. Heath, J. Hodges, J. Fuller)

    1747
    Science Museum, London
    Figure 112

Robertson was at one time master of the Royal Mathematical School at Christ's Hospital, and later headmaster of the Royal Naval Academy, Portsmouth. He then became Clerk, and later Librarian, of the Royal Society. In the dedication to Martin Folkes, President of the Royal Society, Robertson says that his 'work pretends no farther, than to render the use of several mathematical instruments familiar to young students, and all who have occasion for their assistance'. It may therefore have been due to Thomas Heath's prompting that a relatively substantial section was included on the five orders of architecture, which was expanded in the second edition of 1757. The folding plate facing the title page (hand-coloured in this copy) shows Heath's core set of drawing instruments; between the parallel ruler and the sector is the inscription 'Made & contriv'd by Tho Heath in the Strand London'.

61. Thomas Heath
    SET OF DRAWING INSTRUMENTS

    c.1740
    Museum of the History of Science, Oxford, inv. no. 48252
    Figure 114

A 4½-inch pocket set of drawing instruments in a silver-mounted, fish-skin-covered, wooden case. This is just the type of portable set advertised by John Robertson's treatise of 1747 (see catalogue 60). The set includes a combined parallel ruler, protractor, plotting scale, and diagonal scale in silver; a combined sector and square in ivory with silver mounting; a pair of dividers; a drawing pen and a pencil holder, the shaft of which can be unscrewed, revealing a stylus. The most ingenious inclusion is a pair of 'turn-up' compasses with swivel points. These can be configured as dividers, ink compasses, and pencil compasses. The case is signed 'Thos. Heath Fecit.' inside the lid; the parallel ruler and the sector are also signed.

62. George III
    STUDY OF THE IONIC ORDER

    Royal Collection, Windsor K455
    Figure 116

63. George III
    STUDY OF THE IONIC VOLUTE

    Royal Collection, Windsor K445
    Figure 117

64. George III
    COMBINED PLAN AND PERSPECTIVE STUDY OF A VOLUTE

    Royal Collection, Windsor K450
    Figure 118

65. George III
    SKETCH OF PALACE ELEVATION WITH NOTES AND OBSERVATIONS

    Royal Collection, Windsor K1517
    Figure 120

66. George III
    DRAWING FOR A HIGH TOWER OF FIVE STORIES

    Royal Collection, Windsor K1287
    Figure 119

67. George III
    PRELIMINARY SKETCH OF A PALACE FAÇADE

    Royal Collection, Windsor K1548
    Figure 121

68. George III
    PERSPECTIVE VIEW OF A PALACE AND LANDSCAPE

    Royal Collection, Windsor K1540
    Figure 122

George III's architectural education has been studied in Roberts (1987, 57–64), Roberts (1997), Roberts (2004, 93–102), and Watkin (2004). The drawings date from the late 1750s. As has often been noted, the treatises of both Chambers (1759) and Kirby (1761) appear to reflect aspects of the prince's tuition, although it is difficult to pinpoint the respective influence of either author.

69. Joshua Kirby
    The Perspective of Architecture
    (London: for the author)

    1761
    Collection of Howard Dawes
    Figure 123

The primary feature of Kirby's lavish book is the sequence of plates, produced by a number of the most notable engravers of the period. The work exists in variant issues, with differing title pages depending on the presence or absence of a section on Kirby's architectonic sector. Hogarth supplied a frontispiece, whose preparatory chalk drawing is preserved at the British Museum; the sector rests beside a cherub reading Palladio and in the foreground is a new order in honour of George as Prince of Wales. The frontispiece is dated 1760, although the book was not published until April 1761, at the price of £3 3s. Plate LXIV shows a palace whose original drawing at Windsor is signed on the reverse 'GPW' (for George, Prince of Wales, see Watkin (2004, 64)).

70. George Adams
    ARCHITECTONIC SECTOR TO THE DESIGN OF JOSHUA KIRBY

    1760s
    Collection of Howard Dawes
    Figure 124

Brass, in a fitted case; signed by George Adams as maker to the king. Having previously been instrument maker to George as Prince of Wales, Adams was appointed as maker to the new king in December 1760 (Millburn (2000, 94)). This example of the architectonic sector closely matches the form published in Joshua Kirby's *Perspective of Architecture* (1761). One small improvement on the Revesi Bruti pattern introduced by Kirby, or perhaps by Adams, was the readier identification of each line on the proportional arc. Whereas the earlier instrument only identified the line corresponding to each order at the end of the arc, the initials of the orders are here placed on bevelled edges of the legs, so that they are always directly visible against the arc. While brass is the most common material for drawing instruments, and therefore not surprising here, Kirby's text curiously says only that Adams supplied the instrument in silver, ivory, or wood. Parts of two other brass examples are held by the Museum of the History of Science.

71. George Adams
ARCHITECTONIC SECTOR TO THE DESIGN OF JOSHUA KIRBY

1757–60
Royal Institute of British Architects, DRI/122
Figure 125

Silver; signed by Adams as maker to the Prince of Wales. Adams was appointed as instrument maker to the prince at the end of 1756 (Millburn (2000, 75)). Apart from its material, the only significant difference from catalogue 70 is the presence of a retaining strut to preserve the opening of the sector's legs in use. This disappeared from later examples of the instrument and was presumably seen as an unnecessary refinement; Hambly (1988, 137–40) briefly describes and illustrates it.

72. attr. to George Adams
PROTRACTOR FOR QUEEN CHARLOTTE, DESIGNED BY JOSHUA KIRBY

1765
Museum of the History of Science, Oxford, inv. no. 23312
Figure 126

This silver, semicircular protractor carries degrees and a scale of polygons (III–XII) with an index arm on an extended-rule base with a scale of 50 equal parts. The reverse is engraved 'C [crown] R', the royal cipher of Queen Charlotte (1744–1818) and 'Joshua Kirby Invt. Octr. 5. 1765'. G. L'E. Turner (1992) records a second protractor with the same signature; the significance of the date is unknown. Kirby styles himself 'Designer in Perspective to Their Majesties, and Fellow of the Royal and Antiquarian Societies' on the title page to the 1768 edition of Kirby (1754). Previously exhibited in 1996: Bennett and Johnston (1996).

73, 74. attr. to George Adams
VOLUTE COMPASSES, DESIGNED BY DAVID LYLE

1760
Science Museum, London, 1927-1925 and 1927-1924
Figures 127, 128

Described in Morton and Wess (1993, 375–76). Constructed uniformly in silver and steel and each contained in an individual, silver-mounted, shagreen case, they are most probably the work of the royal instrument maker George Adams. Apart from his other mathematical and architectural instruments described here, note the similarity in the shagreen case of a drawing set at the Museum of the History of Science, Oxford, inv. no. 51315.

The larger volute compass is similar in principle to a beam compass, except that the pencil in the slider is not fixed but moves with respect to the stationary point as the compass is turned. A catgut thread connects the end of the helical curved spring to the pencil-holder. As the beam is rotated with the centre fixed, the slider will be moved down the beam at a rate in proportion to the beam's rotation, causing a spiral to be drawn. The various drums (pictured in the foreground of figure 128) give various spirals. The smaller instrument is a simpler, one-handed device, one of three that David Lyle dedicated to George III.

75. George Adams
SILVER MICROSCOPE MADE FOR GEORGE III

c.1763
Museum of the History of Science, Oxford, inv. no. 35086
Figure 129

This elaborate silver microscope, with a fluted Corinthian column as its main support, was made for King George III near the beginning of his reign. When the French astronomer Jerome Lalande visited London in 1763, the microscope was on his itinerary and he noted its cost, material, and optical performance:

> Tuesday 24th May. I lunched at Doctor Pringle's with Bailli de Fleury, etc. We went to see the king's microscope, which cost 3,000 pièces made by Mr Adam. It has all the lenses on a revolving disk and each one has its place; the mounting is of silver. It has a pointer between the oculars which distinguishes movement of a ten-thousandth of an inch to measure magnification. (Millburn (2000, 104))

The monumental base and pillar support a ring of eight objective lenses. These lenses may be used as simple microscopes on one side or as part of a compound system on the other. There are two specimen stages, with positioning screws, and two mirrors to serve the two methods of viewing. The compound body-tube carries a micrometer. The accessories include twelve ivory slides, fish-plate, talc-box, stage forceps, lieberkühn, and simple magnifiers. The mahogany box (not shown) is provided with compartments for all the accessories, and is fitted with a Bramah lock. The pillar stands on a decorated base with feet in the form of figures joined by a transverse bar. An ornate, articulated arm positioned at the top of the pillar holds a lens, and two figures steady the decorated body-tube. The ornamental elements would presumably have required subcontracting with Adams responsible for assembly; the instrument is signed: 'Made by GEORGE ADAMS in Fleet Street LONDON.' A closely similar instrument, once in the possession of the Prince Regent, was transferred from Windsor Castle to the Science Museum, London, in 1949.

76. William Hogarth
'THE FIVE ORDERS OF PERRIWIGS AS THEY WERE WORN AT THE LATE CORONATION, MEASURED ARCHITECTONICALLY', THIRD STATE

1761
Collection of Howard Dawes
Figure 132

Although not his most popular or reproduced print, Hogarth's elaborate proportional dissection of wigs has been regularly analysed: see Stephens and George (1870–1954, iv, no. 3812); Paulson (1989, no. 209); Paulson (1991–93, iii: 349–51). The caricature of James 'Athenian' Stuart as a blockhead on the left is complemented by the advertisement at the foot of the sheet, which takes direct aim at the recently published prospectus for Stuart and Nicholas Revett's *The Antiquities of Athens, Measured and Delineated* (1762). Hogarth mocks the reverence for antiquity and its archaeological measurement in proclaiming

> In about Seventeen Years will be compleated, in Six Volumns, folio, price Fifteen Guineas, the exact Measurements of the Perriwigs of the ancients; taken from the Statues, Bustos & Baso-Relievos, of Athens, Palmira, Balbec, and Rome, by Modesto Perriwig-meter from Lagado. N.B. None will be Sold but to Subscribers.

77. Paul Sandby (?)
'BLOCKS FOR HOGARTH'S WIGS', THIRD STATE

1762
Gainsborough's House
Figure 134

This print takes the blockhead of James Stuart in Hogarth's 'Five Orders of Perriwigs' as one of its starting points. Its contemporary political references have been analysed in detail in Stephens and George (1870–1954, iv, no. 3916) but the identification and significance of the two architectonic sectors have not previously been noticed. They appear on the central figure of the Earl of Bute and on the blockhead of Joshua Kirby on the right.

Catalogue Checklist

78. George Townshend (?)
    'A SCENE OF A PANTAMIME [*sic*] ENTERTAINMENT LATELY EXHIBITED. LONDON: SOLD IN MAY'S BUILDINGS, COVENT GARDEN'

    November 1768
    Museum of the History of Science,
    Oxford, inv. no. 11173
    Figure 135

This etched print in brown ink has been dated to November 1768 and deciphered by Stephens and George (1870–1954, iv, no. 4220) as representing the attack of the remaining members of the Incorporated Society of Artists on those who seceded to form the Royal Academy. The figure of Castor in the background has been identified as William Chambers, announcing 'I have studied abroad' and 'This is a new Advance towards Perfection and Eminence'. Hitherto unnoticed is the sketchy presence of an architectonic sector in his left hand.

Notes

Introduction

1. See Howard Colvin, 'The Practice of Architecture 1600–1840', in *A Biographical Dictionary of British Architects 1600–1840*, 4th edn (New Haven and London: Yale University Press, 2008), 15–37; and John Wilton-Ely, 'The Rise of the Professional Architect in England', in Spiro Kostof, ed., *The Architect: Chapters in the History of the Profession* (New York: Oxford University Press, 1977), 180–208.
2. See Chapters 1 and 2, below.
3. E. G. R. Taylor, *The Mathematical Practitioners of Tudor and Stuart England* (Cambridge: Cambridge University Press, 1954). More recently, see Stephen Johnston, 'Mathematical Practitioners and Instruments in Elizabethan England', *Annals of Science*, 48, no. 4 (1991), 319–344 and 'The Identity of the Mathematical Practitioner in 16th-century England', in Irmgarde Hantsche, ed., *Der 'mathematicus': Zur Entwicklung und Bedeutung einer neuen Berufsgruppe in der Zeit Gerhard Mercators* (Bochum: Brockmeyer, 1996), 93–120.
4. See Rudolf Wittkower, 'English Literature on Architecture', in *Palladio and English Palladianism* (London: Thames and Hudson, 1974), 95–112; Eileen Harris, *British Architectural Books and Writers 1556–1785* (Cambridge: Cambridge University Press, 1990), 41–45, 182–183; and J. A. Bennett, 'Architecture and Mathematical Practice in England, 1550–1650', in John Bold and Edward Chaney, eds., *English Architecture Public and Private: Essays for Kerry Downes* (London: The Hambledon Press, 1993), 23–30.
5. Mark Girouard gives several examples of Elizabethan designer-overseers who also acted as bailiffs, rent collectors, or surveyors. See his *Robert Smythson and the Elizabethan Country House*, 2nd edn (New Haven and London: Yale University Press, 1983), 10–11. Also see Sarah Bendall, Francis Steer, and Peter Eden, *Dictionary of Land Surveyors and Local Map-makers of Great Britain and Ireland, 1530–1850*, 2nd edn, 2 vols. (London: British Library, 1997), which records extant maps from some fifteen masons and master masons born in the sixteenth and seventeenth century.
6. J. A. Bennett, 'The Mechanic's Philosophy and the Mechanical Philosophy', *History of Science*, 24, no. 1 (1986), 1–28. Jürgen Renn and Matteo Valleriani have taken a similar approach in their work on Galileo: see 'Galileo and the Challenge of the Arsenal', *Nuncius*, 16, no. 2 (2001), 481–503.
7. J. A. Bennett, *The Mathematical Science of Christopher Wren* (Cambridge: Cambridge University Press, 1982), 6–13.

Chapter One

1. Sebastiano Serlio, *On Architecture*, ed. Vaughan Hart and Peter Hicks, 2 vols. (New Haven: Yale University Press, 1996–2001), 1.5.
2. For a critical edition of both texts and a useful introduction, see Douglas Knoop, G. P. Jones, and Douglas Hamer, *The Two Earliest Masonic MSS* (Manchester: Manchester University Press, 1938). For an excellent recent interpretation, see Lisa H. Cooper, 'The "Boke of Oure Charges": Constructing Community in the Masons' Constitutions', *Journal of the Early Book Society*, 6 (2003), 1–39. This 'popular' understanding of the figure of Euclid probably derives from Boethius' *Ars geometriae*. For a useful overview, see John Murdoch, 'Euclid: Transmission of the Elements', in Charles Coulston Gillispie, ed., *Dictionary of Scientific Biography* (New York: Scribner, 1970–80), iv.437–459.
3. *The Minor Poems of John Lydgate*, ed. Henry Noble MacCracken, 2 vols. (London: Oxford University Press, 1934), ii.802–808.
4. Robert Recorde, *The Pathway to Knowledg, Containing the First Principles of Geometrie* (London: R. Wolfe, 1551), preface.
5. See Robert Branner, 'Villard de Honnecourt, Reims and the Origin of Gothic Architectural Drawing', *Gazette des Beaux-Arts*, 61 (1963), 129–146. For an exhaustive, recent treatment of the surviving examples in England, see Arnold Pacey, *Medieval Architectural Drawing* (Stroud: Tempus, 2007). On medieval architectural drawing more generally, see L. F. Salzman, *Building in England Down to 1540: A Documentary History* (Oxford: Clarendon Press, 1952), 14–23; and John Harvey, *The Mediaeval Architect* (London: Wayland, 1972), 101–199.
6. See Martin Biddle, 'A Thirteenth-Century Architectural Sketch from the Hospital of St John the Evangelist, Cambridge', *Proceedings of the Cambridge Antiquarian Society*, 54 (1961), 99–108.
7. For the setting-out of the drawing, see Biddle, 'A Thirteenth-Century Architectural Sketch'. For 'constructive geometry', see Lon R. Shelby, 'The Geometrical Knowledge of Medieval Master Masons', *Speculum*, 47, no. 3 (1972), 395–421.
8. On representations of master builders, see Pierre Du Colombier, *Les Chantiers des cathédrales: ouvriers, architectes, sculpteurs*, 2nd edn (Paris: Picard, 1973) and Nicola Coldstream, *Masons and Sculptors* (London: British Museum Press, 1991). For the medieval building site, see Günther Binding, *Der mittelalterliche Baubetrieb in zeitgenössischen Abbildungen* (Stuttgart: Theiss, 2001). For a handy English translation, see Binding, *Medieval Building Techniques*, trans. Alex Cameron (Stroud: Tempus, 2004). For more on the Thorpe album, see Chapter 4.
9. William Worcestre, *Itineraries*, ed. John Harvey (Oxford: Clarendon Press, 1969). Worcestre's notes on Bristol have been edited in *The Topography of Medieval Bristol*, ed. Frances Neale (Bristol: Bristol Record Society, 2000).
10. See Worcestre, *Itineraries*, 314–17, and Worcestre, *The Topography of Medieval Bristol*, 130–131. The sketch is discussed, with a useful elevation of the portal, in Pacey, *Medieval Architectural Drawing*, 118–121. On the Magdalene pattern book, see M. R. James, 'An English Medieval Sketchbook, No. 1916 in the Pepysian Library, Magdalene College, Cambridge', *The Walpole Society*, 13 (1924–25), 1–17.
11. There is a vast literature on medieval architectural drawing. For an orientation, see the magisterial recent edition of the Vienna lodge collection: Johann Josef Böker, ed., *Architektur der Gotik: Gothic Architecture* (Salzburg: Anton Pustet, 2005). On Villard, see Roland Bechmann, *Villard de Honnecourt: la pensée technique au XIIIe siècle et sa communication* (Paris: Picard, 1991). On medieval versus Renaissance drawing conventions, see James S. Ackerman, 'The

Origins of Architectural Drawing in the Middle Ages and Renaissance', in *Origins, Imitation, Conventions: Representation in the Visual Arts* (Cambridge, MA: MIT Press, 2002), 27–65.

12 See Salzman, *Building in England*, 14–23; Lon R. Shelby, 'The Role of the Master Mason in Mediaeval English Building', *Speculum*, 39, no. 3 (1964), 387–403, esp. 390–391; and Pacey, *Medieval Architectural Drawing*, 59–65. On plan construction, see Eric Fernie, 'A Beginner's Guide to the Study of Architectural Proportions and Systems of Length', in Eric Fernie and Paul Crossley, eds., *Medieval Architecture and its Intellectual Context* (London: Hambledon, 1990), 229–237. For examples of reconstructed plans based on measured surveys, see the contributions in Nancy Y. Wu, ed., *Ad Quadratum: The Practical Application of Geometry in Medieval Architecture* (Aldershot: Ashgate, 2002). Böker, however, argues for a stronger role for Gothic plans in both design and construction: *Architektur der Gotik*, 24–27. For a similar view, see Franklin Toker, 'Gothic Architecture by Remote Control: An Illustrated Building Contract of 1340', *Art Bulletin*, 67, no. 1 (1985), 67–95.

13 The drawing was first published by John Harvey in R. A. Skelton and P. D. A. Harvey, eds., *Local Maps and Plans from Medieval England* (Oxford: Clarendon, 1986), 141–146.

14 See William Urry in Skelton and Harvey, *Local Maps and Plans from Medieval England*, 43–58; and Francis Woodman, 'The Waterworks Drawings of the Eadwine Psalter', in Margaret Gibson, T. A. Heslop, and Richard W. Pfaff, eds., *The Eadwine Psalter: Text, Image, and Monastic Culture in Twelfth-century Canterbury* (London and University Park, PA: Modern Humanities Research Association and Pennsylvania State University Press, 1992), 168–177.

15 See M. D. Knowles in Skelton and Harvey, *Local Maps and Plans from Medieval England*, 221–228. The reconstruction is from Bruno Barber and Christopher Thomas, *The London Charterhouse* (London: Museum of London Archaeology Service, 2002), 56.

16 The surveys were published by John H. Harvey, 'Four Fifteenth-century London Plans', *London Topographical Record*, 20 (1952), 1–8.

17 See Phillip Lindley, '"The Singuler Mediacions and Praiers of al the Holie Companie of Heven": Sculptural Functions and Forms in Henry VII's Chapel', in Tim Tatton-Brown and Richard Mortimer, eds., *Westminster Abbey: The Lady Chapel of Henry VII* (Woodbridge: Boydell Press, 2003), 259–293. See the catalogue checklist, below, for additional bibliographic references.

18 The drawing was first published by Mark Girouard, 'Three Gothic Drawings in the Smithson Collection', *Journal of the Royal Institute of British Architects*, 64, no. 1 (1956), 35–36. See the catalogue checklist for further references.

19 See Wolfgang Lefèvre, 'The Emergence of Combined Orthographic Projections', in *Picturing Machines 1400–1700* (Cambridge, MA: MIT Press, 2004), 209–244. On the method of generating elevations, see Lon R. Shelby, ed., *Gothic Design Techniques: The Fifteenth-century Design Booklets of Mathes Roriczer and Hanns Schmuttermayer* (Carbondale, IL: Southern Illinois University Press, 1977).

Chapter Two

1 On the development of the *trace italienne*, see J. R. Hale, 'The Early Development of the Bastion: An Italian Chronology c. 1450–c. 1534,' in *Renaissance War Studies* (London: Hambledon, 1983), 1–29. On the importation of the technique into England, see Hale, 'The Defence of the Realm, 1485–1558', in H. M. Colvin, ed., *The History of the King's Works* (London: HMSO, 1963–82), iv. 367–401; Marcus Merriman, 'Italian Military Engineers in Britain in the 1540s', in Sarah Tyacke, ed., *English Map-Making 1500-1650* (London: British Library, 1983), 57–67; and Andrew Saunders, *Fortress Britain: Artillery Fortification in the British Isles and Ireland* (Liphook: Beaufort, 1989), 48–69.

2 See Merriman, 'Italian Military Engineers in Britain in the 1540s'. On Lee and Rogers, see Lon R. Shelby, *John Rogers: Tudor Military Engineer* (Oxford: Oxford University Press, 1967); John H. Harvey and Arthur Oswald, *English Mediaeval Architects: A Biographical Dictionary down to 1550*, 2nd edn (Gloucester: Alan Sutton, 1987), 175–177, 257–258; Susan Hots in A. W. Skempton and M. M. Chrimes, eds., *Biographical Dictionary of Civil Engineers in Great Britain and Ireland* (London: Thomas Telford, 2002), i. 401–403, 581–584; and the articles by Sarah Bendall and Marcus Merriman in *Oxford Dictionary of National Biography* (Oxford: Oxford University Press, 2004).

3 William Cuningham, *The Cosmographical Glasse* (London: John Day, 1559), 6–7.

4 See Juergen Schulz, 'Jacopo de' Barbari's View of Venice: Map Making, City Views, and Moralized Geography before the Year 1500', *Art Bulletin*, 60, no. 3 (1978), 425–474; Thomas Frangenberg, 'Chorographies of Florence: The Use of City Views and City Plans in the Sixteenth Century', *Imago Mundi*, 46 (1994), 41–64; Lucia Nuti, 'The Perspective Plan in the Sixteenth Century: The Invention of a Representational Language', *Art Bulletin*, 76, no. 1 (1994), 105–128; and David Friedman, '"Fiorenza": Geography and Representation in a Fifteenth-century City View', *Zeitschrift für Kunstgeschichte*, 64, no. 1 (2001), 56–77.

5 Winchester College Muniments, no. 3233. Published and discussed by John Harvey, 'Early Tudor Draughtsmen', in *The Connoisseur Coronation Book*, ed. L. G. G. Ramsey (London: Connoisseur, 1953), 97–102.

6 British Library, Cotton MSS, Augustus I.i.19. See Martin Biddle and John Summerson, 'Dover Harbour', in Colvin, *The History of the King's Works*, iv. 729–768, esp. 732, fn. 4. On other drawings for Dover, see William Minet, 'Some Unpublished Plans of Dover Harbour', *Archaeologia*, 62 (1922), 185–225 and A. Macdonald, 'Plans of Dover in the Sixteenth Century', *Archeologia Cantiana*, 49 (1937), 108–126.

7 See H. M. Colvin, 'Calais', in *The History of the King's Works*, i. 423–456, esp. 444; iii. 337–374, esp. 354–355. The gun towers were never executed. On the dating and attribution of this drawing, see catalogue 10.

8 Both plans are held in the British Library, Cotton MSS, Augustus I, Supp. 14 (attr. to Lee and Rogers) and Augustus I.ii.23 (anonymous Portuguese engineer). See Shelby, *John Rogers*, 5–23; P. D. A. Harvey, 'The Portsmouth Map of 1545 and the Introduction of Scale Maps into England', in John Webb, Nigel Yates, and Sarah Peacock, eds., *Hampshire Studies* (Portsmouth: City Record Office, 1981), 33–49; and P. D. A. Harvey, *Maps in Tudor England* (London: PRO and the British Library, 1993), 31, 36. On the appearance of urban surveying in Italy, see Hilary Ballon and David Friedman, 'Portraying the City in Early Modern Europe: Measurement, Representation, and Planning', in J. B. Harley and David Woodward, eds., *The History of Cartography*, 3 vols (Chicago: University of Chicago Press, 1987–2007), iii, pt1. 680.

9 Shelby, *John Rogers*, 24–35, and Harvey, *Maps in Tudor England*, 31, 40. The fortresses were completed by December 1543 and destroyed in the nineteenth century. See H. M. Colvin, 'Hull, Yorkshire', in *The History of the King's Works*, iv. 472–477.

10 See Harvey, 'The Portsmouth Map of 1545'; Martin Biddle and John Summerson, 'Portsmouth and the Isle of Wight', in Colvin, *The History of the King's Works*, iv. 488–568, esp. 503–506; P. D. A. Harvey, *Maps in Tudor England*, 69, 72–73; and Ballon and Friedman, 'Portraying the City'.

11 For Hull: British Library, Cotton MSS, Augustus I, Supp. 20. For Boulogne: Augustus I.ii.77 and I.ii.53 (pictured). See Shelby, *John Rogers*, 33–34, 98–103. On military surveying generally, see R. A. Skelton, 'The Military Surveyor's Contribution to British Cartography in the 16th Century', *Imago Mundi*, 24 (1970), 77–83.

12 On estate surveys, see E. G. R. Taylor, 'The Surveyor', *Economic History Review*, 17, no. 2 (1947), 121–133; and P. D. A. Harvey, 'Estate Surveyors and the Spread of the Scale-map in England 1550–80', *Landscape History*, 15 (1993), 37–49. On earlier mapping conventions, see

Derek J. Price, 'Medieval Land Surveying and Topographical Maps', *Geographical Journal*, 121, no. 1 (1955), 1–7. For a useful overview, see Peter Barber, 'England I' and 'England II', in David Buisseret, ed., *Monarchs, Ministers, and Maps: The Emergence of Cartography as a Tool of Government in Early Modern Europe* (Chicago: University of Chicago Press, 1992), 26–56, 57–98.

13 On both drawings, see Shelby, *John Rogers*, 34–46.

14 Harvey, 'The Portsmouth Map of 1545', 37. On the earliest scale plans, see P. D. A. Harvey, *The History of Topographical Maps: Symbols, Pictures and Surveys* (London: Thames and Hudson, 1980), 122–132.

Chapter Three

1 National Archives, SP1/171, f. 29r, calendared in J. S. Brewer, J. Gairdner, and R. H. Brodie, eds., *Letters and Papers Foreign and Domestic of the Reign of Henry VIII*, 21 vols in 32 parts (London: Longman, Green, Longman, Roberts & Green, 1864–1920), xvii.234, article 405.

2 Thomas Digges, *Stratioticos* (London, 1579) where Thomas also notes his father's practical work in optics; for Leonard, see Stephen Johnston, 'Leonard Digges', in *Oxford Dictionary of National Biography* (Oxford: Oxford University Press, 2004).

3 Recorde also planned further works beyond those actually published: see Stephen Johnston, 'Robert Recorde', in *Oxford DNB*, with references to earlier literature on Recorde.

4 Eileen Harris, *British Architectural Books and Writers 1556–1785* (Cambridge: Cambridge University Press, 1990), 513–514.

5 Richard More, *The Carpenter's Rule* (London: Felix Kyngston, 1602), sig. A[4]r–v and p. 7; see also Stephen Johnston, 'The Carpenter's Rule: Instruments, Practitioners and Artisans in 16th-century England', in G. Dragoni, A. McConnell and G. L'E. Turner, eds., *Proceedings of the XIth International Scientific Instrument Symposium, Bologna, September 1991* (Bologna: Grafis Edizione, 1994), 39–45. Criticisms echoing those of Leonard Digges can be found in, for example, George Atwell, *The Faithfull Surveyor* (Cambridge: William Nealand, 1662), 116.

6 Richard More, *The Carpenter's Rule*, 2; and John Martyn, *Mensuration Made Easie* (London: Thomas Martyn, 1661), 16–17: many rules 'that are made in the Country, by often transcribing, are become false; yet there is divers made at London, very exact'.

7 Gemini was rewarded with a pension by Henry VIII for copying and publishing Andreas Vesalius' celebrated anatomical plates; for this and his instrument-making, see G. L'E. Turner, *Elizabethan Instrument Makers: The Origins of the London Trade in Precision Instrument Making* (Oxford: Oxford University Press, 2000), 12–20, 34–37.

8 *Tectonicon* (London: Thomas Gemini, 1556), sig. D[5]v.

9 Richard Knight, 'A Carpenter's Rule from the *Mary Rose*', *Tools and Trades*, 6 (1990), 43–55.

10 For all four of these rules, see Turner, *Elizabethan Instrument Makers*, 139–143.

11 Cole supplied elaborate navigational instruments for the Frobisher expeditions: see James McDermott, 'Humphrey Cole and the Frobisher Voyages', in Silke Ackermann, ed., *Humphrey Cole: Mint, Measurement and Maps in Elizabethan England* (London: British Museum, Occasional Paper 126, 1998), 15–19. Other customers included Richard Jugge, Master of the Stationers' Company, and Gabriel Harvey: see Turner, *Elizabethan Instrument Makers*, 23, 112.

12 Their names are mentioned in, for example, the 'Advertisement to the Reader' that follows the preface to Edward Worsop's *A Discouerie of Sundrie Errours and Faults Daily Committed by Lande-meaters, Ignorant of Arithmetike and Geometrie* (London: Gregorie Seton, 1582): see Jim Bennett, '"Braggers that by Showe of their Instrument Win Credit": The *Errours of Edward Worsop*', in Liba Taub and Frances Willmoth, eds., *The Whipple Museum of the History of Science: Instruments and Interpretations* (Cambridge: Whipple Museum, 2006), 79–94.

13 Stephen Johnston, 'Reading Rules: Artefactual Evidence for Mathematics and Craft in Early-Modern England', in Taub and Willmoth, *Whipple Museum*, 252.

14 Howard Colvin, *A Biographical Dictionary of British Architects, 1600–1840*, 4th edn (New Haven and London: Yale University Press, 2008), 1053–1054. Trollope presented pasteboard models to the Corporation of Newcastle, from which their choice was made. For his military engineering, Colvin records a plan of his small fort of 1675 on Lindisfarne Island.

15 John Shute, *The First and Chief Groundes of Architecture* (London: Thomas Marshe, 1563), sig. Aii. v. Quoted in Bennett, 'Architecture and Mathematical Practice,' 27, and see Harris, *British Architectural Books*, 421.

16 Sebastiano Serlio, *The First Booke of Architecture … Entreating of Geometrie* (London: R. Peake, 1611), from the dedication 'to the high and mighty prince Henry, Prince of Wales' (unpaginated).

17 John Summerson, 'Three Elizabethan Architects', *Bulletin of the John Rylands Library*, 40 (1957–58), 202–228. Further references in this section are given only for material not directly cited by Summerson.

18 Cited from Richard Hakluyt's *Principall Navigations* (London, 1589) by John W. Shirley, *Thomas Harriot: A Biography* (Oxford: Clarendon Press, 1983), 102 n.

19 Summerson notes that Adams worked with Cole on an ornamental sundial. Ryther engraved Adams's charts of the defeat of the Armada: see Turner, *Elizabethan Instrument Makers*, 39.

20 For an extended analysis, see Stephen Johnston, 'Making Mathematical Practice: Gentlemen, Practitioners and Artisans in Elizabethan England' (PhD thesis, University of Cambridge, 1994), 212–214; available at http://www.mhs.ox.ac.uk/staff/saj/thesis/practitioners.htm. Also see Frances A. Yates, *Theatre of the World* (London: Routledge & Kegan Paul, 1969), 106–107.

21 John Summerson, 'The Building of Theobalds, 1564–1585', *Archaeologia*, 97, no. 47 (1959), 107–126.

22 For his harbour work at Dover, see National Archives SP12/161/7, SP12/164/12, /13, /17, /25 and SP12/173/27, /27.i. For the harbour project in general, see Martin Biddle and John Summerson, 'Dover Harbour', in Colvin, ed., *The History of the King's Works* (London: HMSO, 1963–82), iv.755–764; Johnston, 'Making Mathematical Practice', ch. 5; and, more recently, Eric H. Ash, *Power, Knowledge, and Expertise in Elizabethan England* (Baltimore: Johns Hopkins University Press, 2004), ch. 2.

23 Johnston, 'Making Mathematical Practice', 251, 255.

24 P. D. A. Harvey, *Maps in Tudor England* (London: PRO and the British Library, 1993), 101.

25 See Girouard, *Robert Smythson and the Elizabethan Country House*, 2nd edn (New Haven and London: Yale University Press, 1983), 168–169, 171.

26 Summerson, 'Three Elizabethan Architects', 223.

27 See John Summerson, 'The Works from 1547–1660', in Colvin, *The History of the King's Works*, iii.126, 133.

28 The form of the hinge and the presence of the thin support plate identify it as a Gunter sector. Edmund Gunter published his design in 1623 but it had been devised and first circulated in about 1606. There are two surviving Gunter sectors in the Science Museum by Charles Whitwell, who died in September 1611. See Turner, *Elizabethan Instrument Makers*, 222–223 and Chapter 7 below.

29 On Simons, see Robert Willis and John Willis Clark, *The Architectural History of the University of Cambridge*, 4 vols. (Cambridge: Cambridge University Press, 1988; 1st edn 1886), ii.248–259, 475–477, 490, 517, 693–694, 736. On the meaning of the term 'architect' and the status of the Elizabethan builder, see Malcolm Airs, *The Tudor & Jacobean Country House: A Building History* (Stroud: Sutton, 1995), 31–56.

Notes

Chapter Four

1. See D. J. Gordon, 'Poet and Architect: The Intellectual Setting of the Quarrel between Ben Jonson and Inigo Jones', *Journal of the Warburg and Courtauld Institutes*, 12 (1949), 152–178; and Christy Anderson, *Inigo Jones and the Classical Tradition* (Cambridge: Cambridge University Press, 2007), 21–48.
2. Gordon, 'Poet and Architect', esp. 163–168. Also see Stephen Orgel and Roy Strong, *Inigo Jones: The Theatre of the Stuart Court*, 2 vols. (London: Sotheby Parke Bernet, Berkeley: University of California Press, 1973), ii.458–459.
3. See John Newman, 'Italian Treatises in Use: The Significance of Inigo Jones's Annotations', in Jean Guillaume, ed., *Les Traités de la Renaissance* (Paris: Picard, 1988), 435–441; Newman, 'Inigo Jones's Architectural Education Before 1614', *Architectural History*, 35 (1992), 18–50; Gordon Higgott, 'Varying with Reason: Inigo Jones's Theory of Design', *Architectural History*, 35 (1992), 51–77; and Anderson, *Inigo Jones*, 88–113. A list of Jones' books held at Worcester can be found in John Harris, Stephen Orgel, and Roy Strong, *The King's Arcadia: Inigo Jones and the Stuart Court* (London: Arts Council of Great Britain, 1973), 217–218.
4. John Harris and Gordon Higgott, *Inigo Jones: Complete Architectural Drawings* (London: Zwemmer, 1989), 126–128.
5. Giacomo Barozzi da Vignola, *Regola delli cinque ordini* (Rome: Andreas Vaccarius, 1607), pl. 30. On the *Regola*, see Christof Thoenes' contributions to the recent exhibition catalogue: Richard J. Tuttle, et al., *Jacopo Barozzi da Vignola* (Milan: Electa, 2002), 350–351. For a useful translation, see Giacomo Barozzi da Vignola, *Canon of the Five Orders of Architecture*, ed. Branko Mitrovic (New York: Acanthus, 1999). See catalogue 29 for an analysis of the construction.
6. Rudolf Wittkower, 'Inigo Jones, Architect and Man of Letters', in *Palladio and English Palladianism* (London: Thames and Hudson, 1974), 51–64. Not all of the dimensions on the elevation were so strictly governed: Gordon Higgott has shown that Jones often manipulated the placement of elements for the sake of visual harmony and optical effect. See his critique of Wittkower's analysis in Harris and Higgott, *Inigo Jones*, 241–243 and, more generally, in Higgott, 'Varying with Reason'.
7. Andrea Palladio, *I quattro libri dell'architettura* (Venice: B. Carampello, 1601), Bk. 4, 40. For a transcription of this page, see Bruce Allsopp, ed., *Inigo Jones on Palladio*, 2 vols. (Newcastle upon Tyne: Oriel, 1970), i.49–50.
8. Buonaiuto Lorini, *Le fortificationi* (Venice: F. Rampazetto, 1609), 16–17. The handwriting in this book has been dated to the 1630s: see Anderson, *Inigo Jones*, 77–82.
9. See Alexander Marr, '"Curious and Useful Buildings": The Mathematical Model of Sir Clement Edmondes', *Bodleian Library Record*, 18, no. 2 (2004), 108–150.
10. On Bodley's donors, see W. D. Macray, *Annals of the Bodleian Library Oxford*, 2nd edn (Oxford: Clarendon, 1890) and Ian Philip, *The Bodleian Library in the Seventeenth and Eighteenth Centuries* (Oxford: Clarendon, 1983). The Benefactor's Register entry is transcribed in Marr, 'Curious and Useful Buildings', 108. Non-book-related gifts are listed in H. H. E. Craster, 'Miscellaneous Donations Recorded in the Benefactor's Register', *Bodleian Quarterly Record*, 4, no. 37 (1923), 22–24. In the early eighteenth-century, the model was kept in front of one of the large windows in the Selden End of the Duke Humphrey Library. See catalogue 32.
11. Euclid, *The Elements of Geometrie*, trans. H. Billingsley (London: John Daye, 1570), 322r. For fold-up models of pyramids, see 314r; for a cut-out dodecahedron, see 321r (misprinted as 341).
12. Both works were executed by craftsmen from Savile's native Yorkshire, who had worked on his family estates. The builders were headed by the mason John Akroyd, who may have previously used the motif of superimposed orders at Howley House, owned by Savile's brother John: see T. W. Hanson, 'Halifax Builders in Oxford', *Transactions of the Halifax Antiquarian Society*, 25 (1928), 253–317; and Catherine Cole, 'The Building of the Tower of Five Orders in the Schools' Quadrangle at Oxford', *Oxoniensia*, 33 (1968), 92–107. For the broader context, see John Newman, 'The Architectural Setting', in Nicholas Tyacke, ed., *The History of the University of Oxford, Volume IV: Seventeenth-century Oxford* (Oxford: Clarendon, 1997), 135–177.
13. John Dee, 'Praeface', in Euclid, *The Elements*, sig. d.iii. The apparent allusion to Kepler is noted in Marr, 'Curious and Useful Buildings', 126. On early modern interest in the Platonic solids, see Margaret Daly Davis, *Piero della Francesca's Mathematical Treatises* (Ravenna: Longo, 1977) and Martin Kemp, 'Geometrical Bodies as Exemplary Forms in Renaissance Space', in Irving Lavin, ed., *World Art: Themes of Unity in Diversity: Acts of the XVIth International Congress of the History of Art* (University Park, PA: Pennsylvania State University Press, 1986), 237–242.
14. Among the other donations that helped to form the Bodleian's initial collection were a good number of treatises on architecture and its allied arts. For a useful compilation, see the appendix to Marr, 'Curious and Useful Buildings'. For the statutes of the Savilian professorships, see Strickland Gibson, ed., *Statuta antiqua Universitatis Oxoniensis* (Oxford: Clarendon, 1931), 528–540, esp. 529.
15. On libraries, see Lucy Gent, *Picture and Poetry 1560–1620: Relations Between Literature and the Visual Arts in the English Renaissance* (Leamington Spa: James Hall, 1981), 66–86; and Malcolm Airs, *The Tudor & Jacobean Country House: A Building History* (Stroud: Sutton, 1995), 37. On book ownership among craftsmen, see Tim Connor, 'The Earliest English Books on Architecture', in John Newman, ed., *Inigo Jones and the Spread of Classicism: Georgian Group Symposium 1986* (1987), 61–68.
16. Henry Peacham, *Graphice*, 2nd edn (London: W. S. for Johne Browne, 1612), 172. The standard reference on Thorpe's life and background remains John Summerson, 'John Thorpe and the Thorpes of Kingscliffe,' repr. in *The Unromantic Castle and Other Essays* (London: Thames and Hudson, 1990), 17–40.
17. See Summerson, 'John Thorpe', 39. For the provenance and contents of the album, see John Summerson, 'The Book of Architecture of John Thorpe in Sir John Soane's Museum', *Walpole Society* 40 (1964–66), 1–119 [whole issue]. For a useful discussion of Thorpe's French sources, see A. A. Tait's review in *The Burlington Magazine*, 108, no. 765 (1966), 634–635.
18. Summerson, 'The Book of Architecture of John Thorpe', 45–46, pl. 5.
19. On the content, significance, and publishing history of Blum's work, see Eileen Harris, *British Architectural Books and Writers 1556–1785* (Cambridge: Cambridge University Press, 1990), 120–122.
20. On the manuscript, see Karl Josef Höltgen, 'An Unknown Manuscript Translation by John Thorpe of Du Cerceau's Perspective', in Edward Chaney and Peter Mack, eds., *England and the Continental Renaissance* (Woodbridge: Boydell, 1990), 215–228; and Christy Anderson, 'The Secrets of Vision in Renaissance England', in Lyle Massey, ed., *The Treatise on Perspective: Published and Unpublished* (New Haven and London: Yale University Press, 2003), 323–347. Thorpe's album contains further perspective studies, based on Serlio: see Summerson, 'The Book of Architecture of John Thorpe', 53–55, pls. 16, 17.

Chapter Five

1. Roger Pratt, *The Architecture of Roger Pratt*, ed. R. T. Gunther (Oxford: Oxford University, 1928), 60.
2. See Howard Colvin, 'What We Mean by Amateur', in Giles Worsley, ed., *The Role of the Amateur Architect: Georgian Group Symposium 1993* (London: The Georgian Group, 1994), 4–6; and Giles Worsley, 'The Gentleman-Professional', 14–20 in the same volume. Also see John Harris and Robert Hradsky, *A Passion for Building: The Amateur Architect in England 1650–1850* (London: Sir John Soane's Museum, 2007).

3. See John Webb, *A Vindication of Stone-Heng Restored*, 2nd edn (London: printed for G. Conyers et al., 1725; 1st edn 1665), 11; and John Evelyn, 'Account of Architects and Architecture', in Roland Fréart de Chambray, *A Parallel of the Antient Architecture with the Modern*, trans. John Evelyn (London: J. Place, 1664), 118.

4. Roger North, *Notes of Me: The Autobiography of Roger North*, ed. Peter Millard (Toronto: University of Toronto Press, 2000), 129–142. See also, Roger North, *Of Building: Roger North's Writings on Architecture*, ed. Howard Colvin and John Newman (Oxford: Clarendon Press, 1981), 4–6.

5. James W. P. Campbell, 'Wren, Architectural Research and the History of Trades in the Early Royal Society', *SVEC*, 6 (2008), 9–27. For a complementary view of the same phenomenon, see Hentie Louw, 'The "Mechanick Artist" in Late Seventeenth-century English and French Architecture', in Michael Cooper and Michael Hunter, eds., *Robert Hooke: Tercentennial Studies* (Aldershot: Ashgate, 2006), 181–199.

6. William Oughtred, *Clavis mathematicae* (Oxford: L. Lichfield, 1652), preface, sig. iii.r.

7. John Evelyn, *The Diary of John Evelyn*, ed. E. S. de Beer, 6 vols. (Oxford: Clarendon, 1955), iii.106, 111 (11, 13 July 1654) and Evelyn, *Sculptura* (London: G. Beedle and T. Collins, 1662), 133.

8. John Ward, *The Lives of the Professors of Gresham College* (London: J. Moore for the Author, 1740), Appendices, 47. Translated in Lena Milman, *Sir Christopher Wren* (London: Duckworth, 1908), 50–51.

9. The dedicatory poem accompanying the letter is dated 1 November 1645. See Christopher Wren (junior), *Parentalia* (Farnborough: Gregg Press, 1965 [reproduction of the 1750 edition of the 'Heirloom' copy at RIBA]), 182 and after 194. For an English translation of the poem, see Milman, *Wren*, 15.

10. See Wren, *Parentalia*, 185; and Milman, *Wren*, 19–20.

11. Wren, *Parentalia*, 187. For caution on this point, see J. A. Bennett, 'A Note on Theories of Respiration and Muscular Action in England c. 1660', *Medical History*, 20 (1976), 59–69.

12. Wren, *Parentalia*, 184. For a translation, see Lisa Jardine, *On a Grander Scale: The Outstanding Career of Sir Christopher Wren* (London: HarperCollins, 2002), 74–75.

13. The calendrical treatise was included in the 1651 edition of Christoph Helwig (Helvicus), *Theatrum Historicum et Chronologicum*. See Wren, *Parentalia*, 195; on the 'diplographical' machine, see 214–216. On the translation, see Oughtred, *Clavis*, preface, sig. iii.r.

14. G. H. Turnbull, 'Samuel Hartlib's Influence on the Early History of the Royal Society', *Notes and Records of the Royal Society*, 10, no. 2 (1953), 114–15. For a fuller list of Wren's early 'Theories, Inventions, Experiments, and Mechanic Improvements', see Wren, *Parentalia*, 198–199.

15. For a recent, influential account, see Jardine, *On a Grander Scale*. For the latest research on this period, see C. S. L. Davies, 'The Youth and Education of Christopher Wren', *English Historical Review*, 123, no. 501 (2008), 300–327.

16. John Summerson, *Sir Christopher Wren* (London: Collins, 1953), 23, 60–61. Summarized in Summerson, 'Christopher Wren: Why Architecture?', repr. in *The Unromantic Castle and Other Essays* (London: Thames and Hudson, 1990), 63–68. For an illustration of Wren's way of thinking, see J. A. Bennett, 'Christopher Wren in Mid-career', in S. J. D. Green and Peregrine Horden, eds., *All Souls Under the Ancien Régime* (Oxford: Oxford University Press, 2007), 76–91. On the danger of overstating the influence of English mathematical practice on Wren's outlook, see Michael Hunter, 'The Making of Christopher Wren', in *Science and the Shape of Orthodoxy: Intellectual Change in Late Seventeenth-Century Britain* (Woodbridge: The Boydell Press, 1995), 45–65.

17. Wren outlined the programme's rationale in 1657 and in 1662. See Wren, *Parentalia*, 202–203, 222–224. For a full reconstruction, see Bennett, *The Mathematical Science of Christopher Wren* (Cambridge: Cambridge University Press, 1982), 82–86. On Robert Hooke's involvement, see Asit K. Biswas, 'The Automatic Rain-Gauge of Sir Christopher Wren, F.R.S.', *Notes and Records of the Royal Society*, 22, no. 1–2 (1967), 94–104; and Jardine, *On a Grander Scale*, 274–276.

18. See Wren, *Parentalia*, 208–209 and Balthasar de Monconys, *Les Voyages*, ed. Charles Henry (Paris: Hermann, 1887 [original edn Lyon, 1665–66]), 61–62. Thomas Birch, *The History of the Royal Society of London*, 4 vols. (London: Millar, 1756–57), i.341, reproduced on pl. III. A later, more complicated version, incorporating some of Hooke's improvements, is described in Nehemiah Grew, *Musaeum Regalis Societatis* (London: For the Author, 1681), 357–358.

19. John Wilkins, *Mathematicall magick* (London: Gellibrand, 1648). Also see Marcus Popplow, 'Why Draw Pictures of Machines? The Social Context of Early Modern Machine Drawings', in Wolfgang Lefèvre, ed., *Picturing Machines 1400–1700* (Cambridge, MA: MIT Press, 2004), 17–48.

20. See Anthony Geraghty, *The Architectural Drawings of Sir Christopher Wren at All Souls College, Oxford: A Complete Catalogue* (Aldershot: Lund Humphries, 2007), 249 (no. 384). Parts of the machine are further explored in Geraghty, nos. 385, 386. For the matching elevation of the All Souls chapel screen, see Geraghty, no. 3.

21. Geraghty, *The Architectural Drawings of Sir Christopher Wren*, 253 (no. 391). The drawing was presented to the Royal Society on 1 February 1665 and was returned to Wren in April the same year. Robert Hooke re-presented it in his *Cometa*, published in Hooke, *Lectures and Collections* (London: J. Martyn, 1678), 40–43. See Bennett, *The Mathematical Science of Christopher Wren*, 65–70. For further discussion, see catalogue 39.

22. See William C. Gibson, 'The Biomedical Pursuits of Christopher Wren', *Medical History*, 14 (1970), 331–341. On the provenance and attribution, see catalogue 40.

23. Geraghty, *The Architectural Drawings of Sir Christopher Wren*, 252 (no. 389, recto and verso).

24. Thomas Willis, *Cerebri anatome cui accessit nervorum descriptio et usus* (London: J. Martyn and J. Allestry, 1664), Preface.

25. Robert Hooke, *Micrographia* (London: J. Martyn and J. Allestry, 1665), Preface.

26. Geraghty, *The Architectural Drawings of Sir Christopher Wren*, 20 (no. 1). For a comprehensive account, see Anthony Geraghty, 'Wren's Preliminary Design for the Sheldonian Theatre', *Architectural History*, 45 (2002), 275–288.

27. Geraghty, *The Architectural Drawings of Sir Christopher Wren*, 46–47 (no. 46), with bibliography. Primary sources for Wren's involvement with the commission are transcribed in A. T. Bolton and H. D. Hendry, eds., *The Wren Society*, 20 vols. (Oxford: Wren Society, 1924–1943), xiii.13–20, 44–45.

28. Geraghty, *The Architectural Drawings of Sir Christopher Wren*, 256 (no. 396). See John Evelyn, *London Revived: Considerations for its Rebuilding in 1666*, ed. E. S. de Beer (Oxford: Clarendon, 1938), 54–55.

Chapter Six

1. The roof is described and illustrated in Robert Plot, *The Natural History of Oxford-shire* (Oxford: Printed at the Theater [sic], 1677), 272–273, pl. 14. For a further explanation, see David Yeomans, *The Trussed Roof: Its History and Development* (Aldershot: Scolar, 1992), 46–49 and James W. P. Campbell, 'Wren and the Development of Structural Carpentry 1660–1710', *Architectural Research Quarterly*, 6, no. 1 (2002), 49–66. Plot compared Wren's truss with John Wallis's 'geometrical flat' roof as examples of ingenious local architecture. This parallel, repeated in Wren, *Parentalia*, 335–337, has often served to confuse the two inventions. As far as we know, Wallis's was never executed, except as a model. See David Yeomans, 'The Serlio Floor and Its Derivations', *Architectural Research Quarterly*, 2, no. 3 (1997), 74–83. On Wren's place in the history of structural

Notes

analysis, see Stanley B. Hamilton, 'The Place of Sir Christopher Wren in the History of Structural Engineering', *Transactions of the Newcomen Society*, 14 (1933–34), 27–42; and Edoardo Benvenuto *An Introduction to the History of Structural Mechanics*, 2 vols. (New York: Springer-Verlag, 1991), ii.313–315.

2. See H. M. Colvin, 'The Building', in David McKitterick, ed., *The Making of the Wren Library: Trinity College, Cambridge* (Cambridge: Cambridge University Press, 1995), 28–49, esp. 40–41. On Duke Humphrey's Library, see Edward Prioleau Warren, 'Sir Christopher Wren's Repair of the Divinity School and Duke Humphrey's Library, Oxford', in Rudolf Dircks, ed., *Sir Christopher Wren A.D. 1632–1723* (London: Hodder and Stoughton, 1923), 233–238; and J. N. L. Myres, 'Recent Discoveries in the Bodleian Library', *Archaeologia*, 101 (1967), 151–168.

3. Margaret Whinney, *Wren* (London: Thames and Hudson, 1971), 118.

4. Gordon Higgott, 'The Revised Design for St Paul's Cathedral, 1685–90: Wren, Hawksmoor and Les Invalides', *The Burlington Magazine*, 146, no. 1217 (2004), 534–547. For an analysis and points of corroboration, see Anthony Geraghty, *The Architectural Drawings of Sir Christopher Wren at All Souls College, Oxford: A Complete Catalogue* (Aldershot: Lund Humphries, 2007), 65–66, 68–69. For a contrasting view regarding the dating of the Revised or Definitive Design, see Kerry Downes, 'Wren, Hawksmoor and Les Invalides Revisited', *The Burlington Magazine*, 150, no. 1261 (2008), 250–252.

5. Wren would have known the building from engravings published in 1683, or perhaps even from original drawings sent to Charles II in November 1678. Further evidence for a link appears in sectional engravings by Pierre Lepautre published as late as 1687. See Higgott, 'The Revised Design', 542, 547. The influence of Les Invalides was first noticed by John Summerson in 'J. H. Mansart, Sir Christopher Wren and the Dome of St Paul's Cathedral', *The Burlington Magazine*, 132, no. 1042 (1990), 32–36.

6. See Higgott, 'The Revised Design', 546, and the same author's contribution to this study, which augments and develops these conclusions. The structural problems revealed by spalling masonry were noticed and described by Jane Lang, *Rebuilding St Paul's after the Great Fire of London* (London: Oxford University Press, 1956), 150–151.

7. For the minutes of the meetings, see Thomas Birch, *The History of the Royal Society of London*, 3 vols. (London: Millar, 1756–57), ii.461 (12 Jan 1671), 464–465 (19 January), 498 (7 December). Hooke's anagram was published in his *A Description of Helioscopes and Some Other Instruments* (London: John Martyn, 1676 [1675]), 31. It was deciphered in Robert Hooke, *The Posthumous Works*, ed. Richard Waller (London: S. Smith and B. Walford, 1705), xxi. More recently, see Jacques Heyman, 'Hooke's Cubico-parabolical Conoid', *Notes and Records of the Royal Society*, 52, no. 1 (1998), 39–50; James W. P. Campbell and Robert Bowles, 'The Construction of the New Cathedral', in Derek Keene, Arthur Burns and Andrew Saint, eds., *St Paul's: The Cathedral Church of London 604–2004* (London: Yale University Press, 2004), 207–219, esp. 215–217; and James W. P. Campbell, *Building St Paul's* (London: Thames and Hudson, 2007), 138–144. For the broader history of the problem, see Clifford A. Truesdell, *The Rational Mechanics of Flexible or Elastic Bodies: 1638–1788, Leonhardi Euleri opera omnia, series secunda, volumina XI, sectio secunda* (Zurich: Orell Füssli, 1960), 64–88.

8. Lydia M. Soo, *Wren's 'Tracts' on Architecture and Other Writings* (Cambridge: Cambridge University Press, 1998), 161.

9. In Rudolf Dircks, ed., *Sir Christopher Wren, A.D. 1632–1723: Bicentenary Memorial Volume* (London: Hodder and Stoughton, 1923), 53; and shortly afterwards in C. Stanley Peach and W. Godfrey Allen, 'The Preservation of St Paul's Cathedral', *Journal of the Royal Institute of British Architects*, 37, no. 18 (1930), 656–676.

10. See Lang, *Rebuilding St Paul's*, 42, 44, 60; and Campbell, *Building St Paul's*, 42–52.

11. On Wren's draughtsmen, see Kerry Downes, *Sir Christopher Wren: The Design of St Paul's Cathedral* (London: Trefoil, 1988), esp. 27–31; Anthony Geraghty, 'Introducing Thomas Laine: Draughtsman to Sir Christopher Wren', *Architectural History*, 42 (1999), 240–245; Geraghty, 'Nicholas Hawksmoor and the Wren City Church Steeples', *Georgian Group Journal*, 10 (2000), 1–14; Geraghty, 'Edward Woodroofe: Sir Christopher Wren's First Draughtsman', *The Burlington Magazine*, 143, no. 1181 (2001), 474–479; and Geraghty, *The Architectural Drawings of Sir Christopher Wren*, 8–13. On mason-contractors, see Douglas Knoop and G. P. Jones, *The London Mason in the Seventeenth Century* (Manchester: Manchester University Press and Quatuor Coronati Lodge, 1935) and Knoop and Jones, 'The Rise of the Mason Contractor', *Journal of the Royal Institute of British Architects*, 43, no. 20 (1936), 1061–1071. On Trinity College, see Colvin, 'The Building', 40.

12. See Downes, *Sir Christopher Wren*, 25, 33–34, 74–77; and Higgott, 'The Revised Design', 540. For precedents of combined views in mason's drawings, see catalogue 10 and 24. Hawksmoor produced further examples of dense, combined views in his drawings for the City church steeples: see Geraghty, 'Nicholas Hawksmoor and the Wren City Church Steeples'.

13. See Downes, *Sir Christopher Wren*, 122–123; and Higgott, 'The Revised Design', 537, n. 24.

14. A. T. Bolton and H. D. Hendry, eds., *The Wren Society*, 20 vols. (Oxford: Wren Society, 1924–1943), xiv.124, 125, 134. See Higgott, 'The Revised Design', 547, n. 86; and Campbell, *Building St Paul's*, 144. Also see Higgott's contribution, below, nn. 26, 28, 31.

Chapter Seven

1. Eileen Harris, *British Architectural Books and Writers 1556–1785* (Cambridge: Cambridge University Press, 1990). The chronological index at p. 513 makes the point particularly clearly.

2. William Leybourn, *The Line of Proportion or Numbers, Commonly Called Gunters Line* (London: G. Sawbridge, 1667), A3r–v.

3. For the sources, composition, and subsequent expansion of the work, see Harris, *British Architectural Books*, 409–411.

4. Vincenzo Scamozzi, *The Mirror of Architecture* (London: William Fisher, 1669), cited from p. 26 of the 1687 edition. Brown's detailed section on timber-framing and roof construction cites William Pope's contribution in Godfrey Richards's translation of Palladio's *First Book of Architecture* (London: G. Richards, 1663). Fisher's compilation may have deliberately echoed Richards's formula of Italian orders and English carpentry practice.

5. For introductions to the history of the sector, see A. J. Turner, *Early Scientific Instruments, Europe 1400–1800* (London: Sotheby's, 1987), 157–160; and Stillman Drake, *Galileo Galilei: Operations of the Geometric and Military Compass, 1606* (Washington, DC: Smithsonian Institution, 1978). The Gunter sector is featured in the portrait of Phineas Pett in Chapter 3 and it is considered in greater detail in Hester K. Higton, 'Elias Allen and the Role of Instruments in Shaping the Mathematical Culture of Seventeenth-century England' (unpublished PhD thesis, University of Cambridge, 1996), ch. 4. Harris, *British Architectural Books*, 127, mistakenly identifies Brown's joint rule as a version of the logarithmic line.

6. Higton, 'Elias Allen', 71–73; and J. A. Bennett, 'Shopping for Instruments in Paris and London', in P. H. Smith and P. Findlen, eds., *Merchants and Marvels: Commerce, Science, and Art in Early Modern Europe* (New York and London: Routledge, 2002), 370–395.

7. Anthony J. Turner, 'Natural Philosophers, Mathematical Practitioners and Timber in Later 17th-Century England', *Nuncius*, 9 (1994), 619–634, esp. 625.

8. Anthony J. Turner, 'Mathematical Instruments and the Education of Gentlemen', *Annals of Science*, 30 (1974), 51–88.

9. Roger North, *Of Building: Roger North's Writings on Architecture*, ed. Howard Colvin and John Newman (Oxford: Clarendon Press, 1981), xvii. Also see Mary Chan's entry in the *Oxford DNB*.

10. Roger North, *Notes of Me: The Autobiography of Roger North*, ed. Peter Millard (Toronto: Toronto University Press, 2000), 100, 92–93, 142.
11. Ibid., 132, 139.
12. Ibid., 134–138.
13. Ibid, 133. Cited in North, *Of Building*, 147–149. North here uses the word 'archetonicall' to describe the sector (British Library Add. MS 32506, f. 60v), although this may be simply a slip; he uses 'archetectonicall' elsewhere in the manuscript: for example, to refer to a desk at f. 64r.
14. North, *Notes of Me*, 129.
15. Ibid., 138.
16. Joseph Moxon, *Mathematicks Made Easy, or, A Mathematical Dictionary* (London: J. Moxon, 1679), 15.
17. Advertisement at the end of the 1701 edition of Moxon's *Mathematicks Made Easy*. For Tuttell, see Hester Higton in *Oxford DNB* and D. J. Bryden, 'A 1701 Dictionary of Mathematical Instruments', in R. G. W. Anderson, J. A. Bennett, and W. F. Ryan, eds., *Making Instruments Count: Essays on Historical Scientific Instruments Presented to Gerard L'Estrange Turner* (Aldershot: Variorum, 1993), 365–382. In producing playing cards, Tuttell was following the Moxons, who published four packs: see D. J. Bryden, 'Capital in the London Publishing Trade: James Moxon's Stock Disposal of 1698, a "Mathematical Lottery"', *The Library*, ser. 6, 19 (1997), 293–350.
18. For Saunders, see the entry in Gloria Clifton, *Directory of British Scientific Instrument Makers, 1550–1851* (London: Zwemmer, 1995).
19. On Heath, see Joyce Brown, *Mathematical Instrument-makers in the Grocers' Company, 1688–1800* (London: Science Museum, 1979), 35, 71–74. Brown places Heath within the extended craft succession of mathematical instrument makers in the Grocers' Company.
20. Brief accounts of each work can be found in Harris, *British Architectural Books*.
21. Keith Maslen, 'Printing for the Author: From the Bowyer Printing Ledgers, 1710–1775', *The Library*, ser. 5, 27 (1972), 302–309.
22. Relevant comparisons can be drawn only with Ottavio Revesi Bruti's 'archisesto' (see p. 124), the 'stylometric rods' published in 1661 by the Leiden-based mathematician Nicolaus Goldmann, and a sector with scales of the orders by Balthasar Neumann (1713). Three surviving sets of Goldmann's rods are known: Whipple Museum of the History of Science, Cambridge, Wh. 0672; Bayerisches Nationalmuseum, Munich, inv. 65619c; and a private collection in the Netherlands (information on the latter from Diederick Wildemann, Nederlands Scheepvaarts-museum). On Goldmann, see Jeroen Goudeau, 'A Northern Scamozzi: Nicolaus Goldmann and the Universal Theory of Architecture', *Annali di Architettura*, 18–19 (2006–07), 235–248. For the Neumann sector, see Christian F. Otto, *Space into Light: The Churches of Balthasar Neumann* (Cambridge, MA: Architectural History Foundation, 1979), 28.
23. See John Ham's preface to his edition of Henry Coggeshall, *The Art of Practical Measuring* (London: Richard King, 1729). Ham's text and diagrams on the 'Scamozzi lines' are based on John Brown's original contribution to *The Mirror of Architecture*, but significantly revised. He provides only the use of the rule for roof calculations and omits Brown's final example of its extension to the five orders.
24. Thomas Carwitham, *The Description and Use of an Architectonick Sector* (London: T. Heath, 1723). On Carwitham, see the exhibition catalogue by Anne Lyles and Robin Hamlyn, *British Watercolours from the Oppé Collection: With a Selection of Drawings and Oil Sketches* (London: Tate Gallery, 1997), 44.
25. Published under the title *A New and Accurate Method of Delineating All the Parts of the Different Orders in Architecture, by Means of a Well Contriv'd, and Most Easily Manag'd Instrument*. Little is known of Malie, though his manuscript survives at Chatsworth: see Harris, *British Architectural Books*, 387.
26. Joshua Kirby, *The Perspective of Architecture* (London: for the author, 1761), 1. See Chapter 8, below, for more on Kirby.
27. William Halfpenny, *A New and Compleat System of Architecture* (London: John Brindley, 1749), 24–25: 'First, chuse the author whose works are most suitable to your purpose (I have made choice for this example Mr Gibb's tuscan cornice)'.
28. John Robertson, *Treatise of Such Mathematical Instruments, As Are Usually Put into a Portable Case* (London: T. Heath, 1747), 107.
29. J. A. Bennett, 'Geometry and Surveying in Early Seventeenth-century England', *Annals of Science*, 48 (1991), 345–354.

Chapter Eight
1. Mike Chrimes, 'Society of Civil Engineers (*act*. 1771–2001)', *Oxford Dictionary of National Biography*, online edn (Oxford University Press, October 2008); http://www.oxforddnb.com/view/theme/93805, accessed 8 January 2009.
2. Edmund Burke, *A Philosophical Enquiry into the Origin of Our Ideas of the Sublime and Beautiful*, 2nd edn (London: Dodsley, 1759; 1st edn 1757), 165–166. For one architect's response to Burke, see Eileen Harris, 'Burke and Chambers on the Sublime and Beautiful', in Douglas Fraser, Howard Hibbard, and Milton J. Lewine, eds., *Essays in the History of Architecture Presented to Rudolf Wittkower on his 65th birthday* (London: Phaidon, 1967), 207–213. More generally, see Rudolf Wittkower, 'Classical Theory and Eighteenth-century Sensibility', in *Palladio and English Palladianism* (London: Thames and Hudson, 1974), 193–204.
3. Allan Ramsay, *The Investigator: Number CCCXXII: To be Continued Occasionally* (London: A. Millar, 1755), 33–34. See Harry Francis Mallgrave, *Modern Architectural Theory: A Historical Survey, 1673–1968* (Cambridge: Cambridge University Press, 2005), 55; and, more generally, Alastair Smart, *Allan Ramsay: Painter, Essayist and Man of the Enlightenment* (New Haven and London: Yale University Press, 1992).
4. For architectural training from the mid-century onward, see the introduction to Giles Worsley, *Architectural Drawings of the Regency Period, 1790–1837* (London: André Deutsch, 1991). For general background, see James Ayres, *Building the Georgian City* (New Haven and London: Yale University Press, 1998), ch. 1.
5. On George's architectural studies, see Jane Roberts, 'Sir William Chambers and George III', in John Harris and Michael Snodin, eds., *Sir William Chambers: Architect to George III* (New Haven and London: Yale University Press, 1997), 41–54; David Watkin, *The Architect King: George III and the Culture of the Enlightenment* (London: Royal Collection, 2004), 64–75; and Jane Roberts, ed., *George III and Queen Charlotte: Patronage, Collecting, and Court Taste* (London: Royal Collection, 2004), 93–107. Three sets of the prince's youthful landscapes and studies are itemized in A. P. Oppé, *English Drawings, Stuart and Georgian Periods, in the collection of His Majesty the King at Windsor Castle* (London: Phaidon, 1950), 19–20. A manuscript inventory of the 'Five Portfolios of Architecture' is preserved in the Royal Library at Windsor: Inventory A, 164/273 and 166/277.
6. Chambers may not have been the exclusive source. Roberts has also noted that the drawing of the Composite order is closer in its details to Joshua Kirby's *Perspective of Architecture* than Chambers' *Treatise*; see Roberts, *George III and Queen Charlotte*, 100.
7. William Robinson, *Proportional Architecture*, 2nd edn (London: W. Dicey, 1736), 47.
8. See Watkin, *The Architect King*, 69; and Roberts, 'Sir William Chambers and George III', 44–45.
9. See the exhibition catalogue *George III: Collector and Patron* (London: Queen's Gallery, Buckingham Palace, 1974) and, more recently, Roberts, ed., *George III and Queen Charlotte*. For an overview of royal patronage across the century, see Jeremy Black, *A Subject for Taste: Culture in Eighteenth-century England* (London: Hambledon and London, 2005), ch. 2.
10. Joshua Kirby, *The Perspective of Architecture* (London: for the author, 1761), 2.

Notes

11 See Oppé, *English Drawings,* 20; and Watkin, *The Architect King,* 64.
12 Royal Archives, Windsor Castle, GEO Add. 32/1742–1760.
13 Kirby, *Perspective of Architecture,* 1.
14 RIBA Collections (held at the Victoria & Albert Museum, London), inv. DRI/122 (silver), catalogue 71 (below); Science Museum, London inv. 1927–1010 (ivory with silver hinge); Royal Artillery Museum, Woolwich, inv. 24/179 (wood and ivory); Howard Dawes collection (brass), catalogue 70 (below); Museum of the History of Science, Oxford (brass, legs only), inv. 76416; and Museum of the History of Science, Oxford (brass, arc only), inv. 40175.
15 G. L'E. Turner, 'The Auction Sales of the Earl of Bute's Instruments, 1793', *Annals of Science,* 23 (1967), 213–242, esp. 234–235. Bute also had 'An architectonic divided brass plate, for constructing the orders, &c. by Sisson'.
16 G. L'E. Turner, 'Queen Charlotte's Protractor and Joshua Kirby', *Bulletin of the Scientific Instrument Society,* 35 (1992), 23–24. Turner records a second protractor with the same signature. Kirby styles himself 'Designer in Perspective to Their Majesties' on the title page of the 1765 edition of his *Dr. Brook Taylor's Method of Perspective Made Easy,* which was dedicated to Bute.
17 Alan Q. Morton and Jane A. Wess, *Public and Private Science: The King George III Collection* (Oxford: Oxford University Press, 1993), 375.
18 See catalogue 75 below for the visit of the French astronomer Jerome Lalande.
19 According to Kirby, the instrument was partly based on new scales 'owing to the ingenuity of that excellent Mr George Adams': Kirby, *The Perspective of Architecture,* 1.
20 Thomas Malton, *An Appendix or Second Part to the Compleat Treatise on Perspective* (London: for the author, 1783), 110–111.
21 Ronald Paulson, *Hogarth's Graphic Works,* 3rd edn (London: The Print Room, 1989), no. 232.
22 Kirby to Hogarth, 3 May 1753, as cited by Felicity Owen, 'Joshua Kirby (1716–74): A Biographical Sketch', *Gainsborough's House Review* (1995–96), 61–76, esp. 64. Compare the more measured language of the work itself: 'Nor would I be thought to desire the Artist to make Use of Scales or Compasses upon all Occasions, and to draw out every Line and Point to a Mathematical Exactness', *Dr. Brook Taylor's Method of Perspective* (Ipswich: for the author, 1754), vii. For the larger context of the perspective dispute, see Amal Asfour and Paul Williamson, 'Splendid Impositions: Gainsborough, Berkeley, Hume', *Eighteenth-Century Studies,* 31 (1998), 403–432.
23 On the print (and Stuart's response), see Frederic George Stephens and Mary Dorothy George, *Catalogue of Political and Personal Satires in the Department of Prints and Drawings in the British Museum,* 11 vols. (London: British Museum, 1870–1954), iv, no. 3812; Paulson, *Hogarth's Graphic Works,* no. 209; Ronald Paulson, *Hogarth,* 3 vols. (New Brunswick, NJ: Rutgers University Press, 1993), iii.349–351; and Jenny Uglow, *Hogarth* (London: Faber and Faber, 1997), 645–647. The sprouting compasses echo Inigo Jones's image of 'Theorica' in figure 57.
24 Stephens and George, *Catalogue of Political and Personal Satires,* iv, no. 3970; Paulson, *Hogarth's Graphic Works,* no. 211. This is the third state of the print in which Pitt is shown directly; previously the figure was that of Henry VIII.
25 For the emblematic character of the Wilkesite prints, see Diana Donald, *The Age of Caricature: Satirical Prints in the Reign of George III* (New Haven and London: Yale University Press, 1996), 50–54. For 'jackboot', see the *Oxford English Dictionary.*
26 For more on the print, see Stephens and George, *Catalogue of Political and Personal Satires,* iv, no. 3916; and David Bindman, *Hogarth and his Times* (London: British Museum, 1997), cat. 116.
27 Jean André Rouquet, *The Present State of the Arts in England* (London: J. Nourse, 1755), 7.

'Geometry and Structure'
by Gordon Higgott

1 From Wren's 'Tract II': see Lydia M. Soo, *Wren's 'Tracts' on Architecture and Other Writings* (Cambridge: Cambridge University Press, 1998), 159. On the British Museum drawing, see Edward Croft-Murray and Paul Hilton, *Catalogue of British Drawings, Volume One: XVI & XVII Centuries* (London: British Museum, 1960), 549–550. See also Teresa Sladen, 'Embellishment and Decoration, 1696–1900', in Derek Keene, Arthur Burns, and Andrew Saint, eds., *St Paul's: The Cathedral Church of London 604–2004* (London: Yale University Press, 2004), 233–256, esp. 235; and James W. P. Campbell, *Building St Paul's* (London: Thames and Hudson, 2007), 142–144.
2 In the last decade or so of his life, to 1735, Hawksmoor's hand became less cursive, more print-like in character, with the bodies of letters formed in detached pen strokes, and with fewer long flourishes: see dated examples reproduced in Vaughan Hart, *Nicholas Hawksmoor: Rebuilding Ancient Wonders* (New Haven and London: Yale University Press, 2002), 223 (1728) and 208 (1734). The other inscription on the British Museum drawing, top right, 'For the Dome of S.t Paul's Church', is in an unidentified early eighteenth-century hand. For Wren's drawing technique in the 1680s and 1690s, and the role of Hawksmoor in this period, see Anthony Geraghty, *The Architectural Drawings of Sir Christopher Wren at All Souls College, Oxford: A Complete Catalogue* (Aldershot: Lund Humphries, 2007), 11–13. The construction history of St Paul's Cathedral can be traced through the works accounts, published in A. T. Bolton and H. D. Hendry, *The Wren Society,* 20 vols. (Oxford: Wren Society, 1924–1943), vols. 13–16; for masons' work at the base of the internal attic in 1690, and at the base of the crossing arches in November 1692, see xv.82 and 105.
3 Jacques Heyman, 'Hooke's Cubico-Parabolical Conoid', *Notes and Records of the Royal Society,* 52, no. 1 (1998), 39–50; Lisa Jardine, *Ingenious Pursuits: Building the Scientific Revolution* (London: Little, Brown and Company, 1999), 72–76; and her *On a Grander Scale: The Outstanding Career of Sir Christopher Wren* (London: HarperCollins, 2002), 420–423. See also Chapter 6, n. 7, above.
4 See chapter 6, nn. 7 and 9.
5 For Poleni's application of the catenary principle at St. Peter's and an analogy with St Paul's, see Jacques Heyman, *The Science of Structural Engineering* (London: Imperial College Press, 1999), 38–42; and the same author's 'Poleni's problem' in *Arches, Vaults and Buttresses: Masonry Structures and Their Engineering* (Aldershot: Variorum, 1996), 345–367. Heyman's *Structural Analysis: A Historical Approach* (Cambridge: Cambridge University Press, 1998), 78–94, explains the history of methods of analysing the mechanics of arches and domes, from Robert Hooke onwards. I am most grateful to Professor Heyman for his advice on my analysis of figure 86, catalogue 45, which I first presented in a lecture to the Sir John Soane's Museum Study Group in October 2002.
6 Gordon Higgott, 'The Revised Design for St Paul's Cathedral, 1685–90: Wren, Hawksmoor and Les Invalides', *The Burlington Magazine,* 146, no. 1217 (2004), 539; the following discussion draws on material in this article. See also Geraghty, *The Architectural Drawings of Sir Christopher Wren,* 66–74; and, for an alternative intepretation of the origins of the south elevation in figure 137, Kerry Downes, 'Wren, Hawksmoor and Les Invalides Revisited', *The Burlington Magazine,* 150, no. 1261 (2008), 250–252.
7 John Summerson, 'The Penultimate Design for St Paul's Cathedral', *The Burlington Magazine,* 103, no. 696 (1961), 83–89, republished with amendments in Summerson, *The Unromantic Castle and Other Essays* (London: Thames and Hudson, 1990), 69–78. I have argued that the Penultimate design was not superseded when work began, but remained current until 1685, a possibility that Summerson also considered: Higgott, 'The Revised Design', 537–539. See also Geraghty, *The Architectural Drawings of Sir Christopher Wren,* 65.
8 Geraghty, *The Architectural Drawings of Sir Christopher Wren,* nos. 66–75; Summerson saw the design in figure 143 as the direct successor to All Souls II.17 (Geraghty no. 68), where the

9. L. J. de Boulencourt, *Description générale de l'hostel royale des Invalides établi par Louis le Grand dans la Plaine de la Grenelle prés Paris* (Paris: chez l'auteur, 1683), with engravings by Jean and Daniel Marot and Pierre and Jean Lepautre.
10. London Metropolitan Archives, St Paul's Collection; Kerry Downes, *Sir Christopher Wren: The Design of St Paul's Cathedral* (London: Trefoil, 1988), 115 (no. 94). For the dating of this engraving, see Higgott, 'The Revised Design', 544–546. Inscribed by Hawksmoor, it bears his sketched alternatives for the cornice profiles, which correspond to those on the section in figure 149. It was previously thought to represent the 1675 design, along with figure 137 and about fifteen studies for the dome. See Downes, *Sir Christopher Wren*, 23–26 and 115–116.
11. See Hulsbergh's engraved elevation of 1726, reproduced in Bolton and Hendry, *The Wren Society*, xiv. pl. II and III and the section of the Greek Cross design (the precursor of the Great Model) in Geraghty, *The Architectural Drawings of Sir Christopher Wren*, no. 54.
12. Published individually by G. G. de Rossi in 1687, with a dedication to 'Giovan Battista Abbate del Palagio' and inscribed beneath the title as having been 'newly measured, drawn and engraved by Alessandro Specchi'. Specchi went on to supply many plates for Carlo Fontana's *Templum Vaticanum* (Rome, 1694), but this remarkable image was not among them. It seems never to have been published in book form but, like the engravings of Les Invalides, must have quickly reached Wren through one of the many stationers and booksellers in the vicinity of St Paul's (for which, see Anthony Geraghty, 'Robert Hooke's Collection of Architectural Books and Prints', *Architectural History*, 47 (2004), 113–125).
13. From a little-known suite of fourteen large prints, issued as work was about to begin on the construction of the Invalides dome above the roofline, and partly intended for contractors bidding for the work: see Higgott, 'The Revised Design', 544.
14. Higgott, 'The Revised Design', 545–546. The previous attribution to Wren's draughtsman Woodroofe, who died in November 1675, was a vital plank in the argument supporting a 1675 date for the Revised design, predating the start of masonry construction in August 1675; see Kerry Downes, 'Sir Christopher Wren, Edward Woodroffe, J. H. Mansart and architectural history', *Architectural History*, 37 (1994), 37–67. Now accepting Hawksmoor's authorship, Downes recently suggested a c.1699 date for the drawing, presumably by association with the authorized engravings (figure 153) prepared soon after: see Downes, 'Wren, Hawksmoor and Les Invalides Revisited', 250). In this case, however, the handwriting and techniques of the drawing are closely datable, c.1690: see Geraghty, *The Architectural Drawings of Sir Christopher Wren*, 13.
15. See Godfrey Allen's section of the pier at crypt and church floor levels, in Bolton and Hendry, *The Wren Society*, xvi. pl. 4; and Stanley B. Hamilton, 'The Place of Sir Christopher Wren in the History of Structural Engineering', *Transactions of the Newcomen Society*, 14 (1933–34), 27–42.
16. See Jane Lang, *Rebuilding St Paul's after the Great Fire of London* (Oxford: Oxford University Press, 1956), 151–122; and references in the works accounts to masons opening joints in the 'lower work' from 1689 onwards: Bolton and Hendry, *The Wren Society*, xiv. 80, 89, 98.
17. The soffit bands are first mentioned as 'Ribbs' of the 'Great Arch' in Edward Strong's bill of November 1692, cited in note 2.
18. The restoration work of 1925–30 is described in Martin Stancliffe, 'The Conservation of the Fabric', in Keene, Burns, and Saint, *St Paul's*, 299–300. The wooden model survives at St Paul's and is illustrated in Bolton and Hendry, *The Wren Society*, xvi. pl. 6.
19. The 3 : 4 proportional relationship of inner dome to crossing square can be traced back to the plan of the Warrant design at All Souls (Geraghty no. 71), where the nave is 120 feet wide externally, and the crossing square and west end are both 160 feet wide; and to the quarter plan of the Penultimate design (figure 142), in which the nave is still 120 feet wide externally (compared with 123 feet in the fabric), and this same dimension is adopted, for the first time, for the internal octagon to the backs of the soffit bands.
20. See Chapter 6, n. 7.
21. John Evelyn, *The Diary of John Evelyn*, ed. E. S. de Beer, 6 vols. (Oxford: Clarendon, 1955), iii. 598 (6 or 7 December 1671).
22. See Chapter 6, n. 7. The deciphered Latin text reads, *Ut pendet continuum flexile, sic stabit contiguum rigidum inversum*. A closer, more literal translation is 'As it hangs in a continuous flexible form, so it will stand contiguously rigid when inverted'; see Heyman, 'Hooke's cubico-parabolical conoid', 40; and Jardine, *Ingenious Pursuits*, 73–74.
23. H. W. Robinson and W. Adams, *The Diary of Robert Hooke, 1662–1680* (London: Taylor and Francis, 1935), 163. The previous phrase, 'He promised Fitch at Paules', associates the design in question with St Paul's Cathedral (the word 'model' signifying either a 'design' or a built model in the seventeenth century). It is not known how Wren could then have applied Hooke's 'principle' to the design of St Paul's, but one possibility is that he used the catenary arch in the vaults in the crypt, for, in a study drawing of exactly this period, Wren uses a catenary profile for the vaults of the window apertures; see Downes, *Sir Christopher Wren*, 88 (no. 60).
24. David Gregory's 'Catenaria' was first published in *Philosophical Transactions of the Royal Society*, 19, no. 231 (1697), 637–652. The above excerpt is from the translation by Samuel Ware, *A Treatise of the Properties of Arches, and their Abutment Piers* (London: J. Taylor, 1809), as discussed in Heyman, *The Science of Structural Engineering*, 29–30, where he notes that, in 1691, three mathematicians – Christiaan Huygens, Johann Bernoulli, and Gottfried Leibniz – solved the equation of the catenary independently of each other, but without publishing the results in full. See also John Stillwell, *Mathematics and its History* (New York: Springer, 2004), 236–237.
25. William Dickinson (c.1671–1725), the draughtsman of figure 88, began work in the St Paul's office as a measurer in 1696 (see retrospective payment in December 1700: Bolton and Hendry, *The Wren Society*, xv. 69). His hand is less fluent than Hawksmoor's, and in his annotations he habitually used colons to separate feet and inches. His earliest identifiable drawings at St Paul's are for the peristyle in c.1696–1700, when the dimensions were fixed but the details were not fully resolved; see Downes, *Sir Christopher Wren*, 154 (no. 161).
26. Strong's three or four small-scale models of the dome, discussed below, have all disappeared, but as they were all prepared in the early phase, up to 1695, and are of segments of the plan rather than its whole circumference, they are likely to have represented the structure below the curvature of the inner dome. A surviving example of a small-scale masonry model is that of the dome of James Gibbs's Radcliffe Camera in Oxford (1737–48); see Howard Colvin, *Unbuilt Oxford* (New Haven and London: Yale University Press, 1983), 73, fig. 81. Strong was Wren's leading master mason. See Campbell, *Building St Paul's*, 75–76 and pl. 7, and the discussion in catalogue 48, below.
27. The work of Tijou and that of other ironsmiths at St Paul's is listed in Bolton and Hendry, *The Wren Society*, xv. xxxiii–xxxiv; he was paid for 'the Great Iron Chain or Girdle round the Dome' in March 1706 (p. 133).
28. Payments for 'Modell for 1/4 part of the Dome', February 1691; and for 'making part of a Modell in small Stones for part of the Dome and a little Modell for the vaulting of the Choire'; May 1691: Guildhall Library, MS 25, 471/32, ff. 21, 40, transcribed in Bolton and Hendry, *The Wren Society*, xiv. 80, 86. The drawings can be identified in Downes, *Sir Christopher Wren* as nos. 87–88, 89, 90–91, 97 (figure 151), 98, 99 (figure 87, catalogue 46). Four part-plans of these twenty-four-sided

Notes

domes are drawn over a half internal plan of the cathedral that was prepared for the proof engraving of the long section (figure 145) (Downes, *Sir Christopher Wren*, no. 94). This indicates that the twenty-four-sided dome was the immediate successor to the sixteen-sided dome of the Revised design.

29 Wren was responding to instructions from the new building commission of William and Mary, which did not formally commence business until 1692, three years after their accession. See the commission minutes for 7 July 1692 and 2 October 1693 in Bolton and Hendry, *The Wren Society*, xvi.70–71, 75.

30 See Anthony Geraghty, 'Nicholas Hawksmoor and the Wren City Church Steeples', *Georgian Group Journal*, 10 (2000), 3–4 and Geraghty, *The Architectural Drawings of Sir Christopher Wren*, 13.

31 Payment to Edward Strong of £2 15s 'for 18 days work of a Mason in making a large Modell of 1/8th of the Great Dome at 2s 8d per day; for 3 days work of a Carver about the same': Guildhall Library, MS 25, 471/35, f. 51, transcribed in Bolton and Hendry, *The Wren Society*, xiv.134. The latter part of the payment would suggest that the model included the peristyle, as in Hawksmoor's plan. On the sketch, the inner wall of the drum is inscribed '5', confirming a scale of twenty feet to an inch; the section of the lower drum is three-quarters of an inch wide on the drawing. Significantly, the inscribed dimensions on the plan for the depths of the peristyle and inner wall (15 feet 4 inches and 5 feet) are too small by the scale (17 feet and 6 feet respectively); they correspond instead to the dimensions on the sketch in figure 152, suggesting an ad hoc process, whereby the design is not that illustrated in the drawing, but one stage beyond.

32 See the payment to Edward Strong for construction of the lower drum up to the base of the peristyle in July–September 1700, its inside circular wall 'inclining to the Centre of the Dome 1 inch in a foot'. Bolton and Hendry, *The Wren Society*, xv.63–64.

33 Guildhall Library, MS 25, 471/37, f. 20, transcribed in Bolton and Hendry, *The Wren Society*, xv.5: payment of 15 shillings for 6 days' work by one mason, with no carved work.

34 Bolton and Hendry, *The Wren Society*, xv.6; the first payments for 'freestone Work on the Inside of the Dome' were in March 1696 (p. 7). Sloping the inner wall eased the task of centring and constructing the brickwork. This was done from timber platforms, cantilevered inwards from the perimeter, and not built up from the floor below. One of these platforms is sketched in pencil on the left side of figure 88, catalogue 47.

35 C. Peach and W. Godfrey Allen, 'The Preservation of St Paul's Cathedral', *Journal of the Royal Institute of British Architects*, 37, no. 18 (1930), 661.

36 See above, n. 25. The designs for the peristyle and lantern, and their relationship to the sequence of work on the fabric, are described in Downes, *Sir Christopher Wren*, nos. 157–165. It should be noted, however, that figure 93, catalogue 50 (Downes, no. 159) mistakenly shows the entablature of the dome too steeply raked and the soffits of the arch misplaced in relation to the structure above.

37 Engravings of the plan and north elevation (by Jan Kip) and the west front (by Simon Gribelin) were based on designs approved by the building commission in February 1700 (Bolton and Hendry, *The Wren Society*, xiv.pls. 6, 10, 12; and xvi.98). Payments for Kip's two engravings in June 1701 (xv.71; Gribelin's followed in May 1702: xv.84), date the design on the right-hand side of Dickinson's drawing to c.1700–01.

38 The height of the inner dome in the left-hand section is 224 feet above church floor level. This is double the width of the inner drum at the Whispering Gallery (112 feet) and approximately the height of the inner dome as built to the centre of the oculus. In a final, unrecorded revision before the start of work at the base of the inner dome and brick cone in 1705, the hemispherical curve of the inner drum in figure 88 was made flatter to merge more effectively with the sloping line of the wall below.

39 See Campbell, *Building St Paul's*, 67–69; and Lang, *Rebuilding St Paul's*, 220. A detailed report by Wren on 23 March 1697 explains to the building commission the nature and consequences of the reduction in funding at that time: Bolton and Hendry, *The Wren Society*, xvi.82–84.

40 For the completion of the lantern on 26 October 1708, see Campbell, *Building St Paul's*, 6–7. External cladding, and other works associated with the dome, continued until 1710.

41 See Peach and Allen, 'The Preservation of St Paul's Cathedral'. Also see Martin Stancliffe, 'Conservation of the Fabric', in Keene, Burns and Saint, *St Paul's*, 298–300; and Campbell and Bowles, 'The Construction of the New Cathedral', in ibid., 215–217. A valuable early account of the condition of the fabric is Mervyn Macartney's 'The Present Condition of St Paul's Cathedral', *Journal of the Royal Institute of British Architects*, 15 (1907), 53–71. Dickinson's drawing was not discovered until 1935, and so its implications for understanding the structural behaviour of Wren's dome were unknown to the architects and engineers who planned the strengthening works in the 1920s; see A. T. Bolton's 1936 appraisal of the drawing in *The Wren Society*, xiii.xv.

Bibliography

Ackerman, James S. (2002). 'The Origins of Architectural Drawing in the Middle Ages and Renaissance'. In *Origins, Imitation, Conventions: Representation in the Visual Arts*, 27–65. Cambridge, Mass.: MIT Press.

Ackermann, Silke (1998) (ed.). *Humphrey Cole: Mint, Measurement and Maps in Elizabethan England*. British Museum Occasional Paper 126. London: British Museum.

Airs, Malcolm (1995). *The Tudor & Jacobean Country House: A Building History*. Stroud: Sutton.

Allsopp, Bruce (1970) (ed.). *Inigo Jones on Palladio*. 2 vols. Newcastle upon Tyne: Oriel.

Anderson, Christy (2003). 'The Secrets of Vision in Renaissance England'. In *The Treatise on Perspective: Published and Unpublished*, ed. Lyle Massey, 323–47. New Haven and London: Yale University Press.

—— (2007). *Inigo Jones and the Classical Tradition*. Cambridge: Cambridge University Press.

Asfour, Amal, and Williamson, Paul. 'Splendid Impositions: Gainsborough, Berkeley, Hume'. *Eighteenth-Century Studies*, 31, 403–432.

Ash, Eric H. (2004). *Power, Knowledge, and Expertise in Elizabethan England*. Baltimore: The Johns Hopkins University Press.

Atwell, George (1662). *The Faithfull Surveyor*. Cambridge: William Nealand.

Ayres, James (1998). *Building the Georgian City*. New Haven and London: Yale University Press.

Babington, Charles Cardale (1874). *History of the Infirmary and Chapel of the Hospital and College of St John the Evangelist at Cambridge*. Cambridge: Deighton, Bell, and Co.

Ballon, Hilary, and Friedman, David (2007). 'Portraying the City in Early Modern Europe: Measurement, Representation, and Planning'. In *The History of Cartography*, Volume 3: *Cartography in the European Renaissance*, ed. J. B. Harley and David Woodward, pt 1, 680–704. Chicago: University of Chicago Press.

Barber, Bruno, and Thomas, Christopher (2002). *The London Charterhouse*. London: Museum of London Archaeology Service.

Barber, Peter (1992). 'England I: Pageantry, Defense, and Government: Maps at Court to 1550' and 'England II: Monarchs, Ministers, and Maps, 1550–1625'. In *Monarchs, Ministers, and Maps: The Emergence of Cartography as a Tool of Government in Early Modern Europe*, ed. David Buisseret, 26–56, 57–98. Chicago: University of Chicago Press.

Bechmann, Roland (1991). *Villard de Honnecourt: La Pensée technique au XIIIe siècle et sa communication*. Paris: Picard.

Bendall, Sarah, Francis Steer, and Peter Eden (1997) (eds.). *Dictionary of Land Surveyors and Local Map-makers of Great Britain and Ireland, 1530–1850*. 2nd edn. 2 vols. London: British Library.

Bennett, J. A. (1976). 'A Note on Theories of Respiration and Muscular Action in England c. 1660'. *Medical History*, 20, 59–69.

—— (1982). *The Mathematical Science of Christopher Wren*. Cambridge: Cambridge University Press.

—— (1986). 'The Mechanic's Philosophy and the Mechanical Philosophy'. *History of Science*, 24, no. 1, 1–28.

—— (1993). 'Architecture and Mathematical Practice in England, 1550–1650'. In *English Architecture Public and Private: Essays for Kerry Downes*, ed. John Bold and Edward Chaney, 23–30. London: The Hambledon Press.

—— (2002). 'Shopping for Instruments in Paris and London'. In *Merchants and Marvels: Commerce, Science, and Art in Early Modern Europe*, ed. Pamela H. Smith and Paula Findlen, 370–95. New York and London: Routledge.

—— (2006). '"Braggers That by Showe of Their Instrument Win Credit": The Errours of Edward Worsop'. In Taub and Willmoth (2006), 79–94.

—— (2007). 'Christopher Wren in Mid-career'. In *All Souls under the Ancien Régime*, ed. S. J. D. Green and Peregrine Horden, 76–91. Oxford: Oxford University Press.

Bennett, Jim, and Johnston, Stephen (1996). *The Geometry of War, 1500–1750*. Oxford: Museum of the History of Science.

Benvenuto, Edoardo (1991). *An Introduction to the History of Structural Mechanics*. 2 vols. New York: Springer-Verlag.

Biddle, Martin (1961). 'A Thirteenth-century Architectural Sketch from the Hospital of St John the Evangelist, Cambridge'. *Proceedings of the Cambridge Antiquarian Society*, 54, 99–108.

Biddle, Martin, and Summerson, John (1982a). 'Dover Harbour'. In Colvin (1963–82), iv.729–68.

—— (1982b). 'Portsmouth and the Isle of Wight'. In Colvin (1983–82), iv.488–568.

Binding, Günther (2001). *Der mittelalterliche Baubetrieb in zeitgenössischen Abbildungen*. Stuttgart: Theiss.

—— (2004). *Medieval Building Techniques*, tr. Alex Cameron. Stroud: Tempus.

Bindman, David (1997). *Hogarth and His Times*. London: British Museum.

Birch, Thomas (1756–57). *The History of the Royal Society of London*. 4 vols. London: Millar.

Birch, Thomas (1737) (ed.). *Miscellaneous Works of Mr. John Greaves*. 2 vols. London: J. Brindley and C. Corbett.

Biswas, Asit K. (1967). 'The Automatic Rain-gauge of Sir Christopher Wren, F.R.S.'. *Notes and Records of the Royal Society*, 22, no. 1–2, 94–104.

Black, Jeremy (2005). *A Subject for Taste: Culture in Eighteenth-century England*. London: Hambledon and London.

Blum [Bloome], Hans (1601). *The Booke of Five Collumnes of Architecture*. London: H. Wounteel.

Böker, Johann Josef (2005) (ed.). *Architektur der Gotik: Gothic Architecture*. Salzburg: Anton Pustet.

Bolton, A. T., and Hendry, H. D. (1924–43) (eds.). *The Wren Society*. 20 vols. Oxford: Wren Society.

Boulencourt, L. J. de (1683). *Description générale de l'Hostel royale des Invalides établi par Louis le Grand dans la Plaine de la Grenelle prés Paris*. Paris: chez l'auteur.

Branner, Robert (1963). 'Villard de Honnecourt, Reims and the Origin of Gothic Architectural Drawing'. *Gazette des Beaux-Arts*, 61, 129–46.

Brewer, J. S., Gairdner, J., and Brodie, R. H. (1862–1932). *Letters and Papers Foreign and Domestic of the Reign of Henry VIII*. 23 vols in 35 parts. London: Longman.

Brown, John (1661). *The Description and Use of a Joynt-rule*. London: for J. Brown and H. Sutton.

Brown, Joyce (1979). *Mathematical Instrument-makers in the Grocers' Company, 1688–1800*. London: Science Museum.

Bruti, Ottavio Revesi Bruti (1627). *Archisesto per formar con facilità li cinque ordini d'Architettura*. Vicenza: Amadio.

Bibliography

Bryden, D. J. (1993). 'A 1701 Dictionary of Mathematical Instruments'. In *Making Instruments Count: Essays on Historical Scientific Instruments Presented to Gerard L'Estrange Turner*, ed. R. G. W. Anderson, J. A. Bennett, and W. F. Ryan, 365–82. Aldershot: Variorum.

—— (1997). 'Capital in the London Publishing Trade: James Moxon's Stock Disposal of 1698, a "Mathematical Lottery"'. *The Library*, ser. 6, 19, 293–350.

—— (1998). 'A Fortification Sector to the 1673 Design of Sir Jonas Moore'. *Antiquaries Journal*, 78, 323–43.

Burke, Edmund (1759). *A Philosophical Enquiry into the Origin of Our Ideas of the Sublime and Beautiful*. 2nd edn. London: Dodsley. (First edn 1757.)

Camerota, Filippo (2004). 'Renaissance Descriptive Geometry'. In Lefèvre (2004b), 175–208.

Campbell, James W. P. (2002). 'Wren and the Development of Structural Carpentry 1660–1710'. *Architectural Research Quarterly*, 6, no. 1, 49–66.

—— (2007). *Building St Paul's*. London: Thames and Hudson.

—— (2008). 'Wren, Architectural Research and the History of Trades in the Early Royal Society'. *SVEC*, 6, 9–27.

Campbell, James W. P., and Bowles, Robert. 'The Construction of the New Cathedral'. In Keene et al. (2004), 207–19.

Carpo, Mario (2004). 'Drawing with Numbers: Geometry and Numeracy in Early Modern Architectural Design'. *Journal of the Society of Architectural Historians*, 62, no. 4, 448–469.

Carwitham, Thomas (1723). *The Description and Use of an Architectonick Sector*. London: T. Heath.

Chambers, William (1759). *A Treatise on Civil Architecture*. London: for the author.

Chambray, Roland Fréart de (1664). *A Parallel of the Antient Architecture with the Modern*, tr. John Evelyn. London: by Thomas Roycroft for John Place.

Chan, Mary (1995) (ed.). *The Life of the Lord Keeper North by Roger North*. Studies in British History, 41. Lewiston, NY: Edwin Mellen Press.

Clifton, Gloria (1995). *Directory of British Scientific Instrument Makers, 1550–1851*. London: Zwemmer.

Coggeshall, Henry (1729). *The Art of Practical Measuring*. London: Richard King.

Coldstream, Nicola (1991). *Masons and Sculptors*. London: British Museum Press.

Cole, Catherine (1968). 'The Building of the Tower of Five Orders in the Schools' Quadrangle at Oxford'. *Oxoniensia*, 33, 92–107.

Colvin, H. M. (1983). *Unbuilt Oxford*. New Haven and London: Yale University Press.

—— (1994). 'What We Mean by Amateur'. In *The Role of the Amateur Architect: Georgian Group Symposium 1993*, ed. Giles Worsley, 4–6. London: The Georgian Group.

—— (1995). 'The Building'. In *The Making of the Wren Library: Trinity College, Cambridge*, ed. David McKitterick, 28–49. Cambridge: Cambridge University Press.

—— (2008). *A Biographical Dictionary of British Architects 1600–1840*. 4th edn. New Haven and London: Yale University Press.

Colvin, H. M. (1963–82) (ed.). *The History of the King's Works*. 6 vols. London: HMSO.

Connor, Tim (1987). 'The Earliest English Books on Architecture'. In *Inigo Jones and the Spread of Classicism: Georgian Group Symposium 1986*, ed. John Newman, 61–8. London: The Georgian Group.

Cooper, Lisa H. (2003). 'The "Boke of Oure Charges": Constructing Community in the Masons' Constitutions'. *Journal of the Early Book Society*, 6, 1–39.

Coulton, G. G. (1928). *Art and the Reformation*. Oxford: Blackwell.

Craster, H. H. E. (1923). 'Miscellaneous Donations Recorded in the Benefactors' Register'. *Bodleian Quarterly Record*, 4, no. 37, 22–4.

Croft-Murray, Edward, and Hilton, Paul (1960–63). *Catalogue of British Drawings*. London: British Museum.

Cuningham, William (1559). *The Cosmographical Glasse*. London: John Day.

Dallaway, James (1834). *Antiquities of Bristow in the Middle Centuries*. Bristol: Mirror Office.

Davies, C. S. L. (2008). 'The Youth and Education of Christopher Wren'. *English Historical Review*, 123, no. 501, 300–327.

Davis, Margaret Daly (1977). *Piero della Francesca's Mathematical Treatises*. Ravenna: Longo.

Delano-Smith, Catherine, and Kain, Roger J. P. (1999). *English Maps: A History*. London: British Library.

Digges, Leonard (1556). *A Boke Named Tectonicon*. London: Thomas Gemini.

—— (1571). *A Geometrical Practise, Named Pantometria*. London.

Digges, Thomas (1579). *Stratioticos*. London.

Dircks, Rudolf (1923) (ed.). *Sir Christopher Wren A.D. 1632–1723: Bicentenary Memorial Volume*. London: Hodder and Stoughton.

Donald, Diana (1996). *The Age of Caricature: Satirical Prints in the Reign of George III*. New Haven and London: Yale University Press.

Downes, Kerry (1982). *Sir Christopher Wren: An Exhibition Selected by Kerry Downes at the Whitechapel Art Gallery*. London: Whitechapel Art Gallery.

—— (1988). *Sir Christopher Wren: The Design of St Paul's Cathedral*. London: Trefoil.

—— (1991). *Sir Christopher Wren and the Making of St Paul's*. London: Royal Academy of the Arts.

—— (1994). 'Sir Christopher Wren, Edward Woodroffe, J. H. Mansart and Architectural History'. *Architectural History*, 37, 37–67.

—— (2008). 'Wren, Hawksmoor and Les Invalides Revisited'. *Burlington Magazine*, 150, no. 1261, 250–52.

Drake, Stillman (1978). *Galileo Galilei: Operations of the Geometric and Military Compass, 1606*. Washington, DC: Smithsonian Institution.

du Colombier, Pierre (1973). *Les Chantiers des cathédrales: ouvriers, architectes, sculpteurs*. 2nd edn Paris: Picard.

Dürer, Albrecht (1525). *Underweysung Der Messung*. Nuremberg.

Euclid (1570). *The Elements of Geometrie*, tr. H. Billingsley. London: John Daye.

Evelyn, John (1662). *Sculptura*. London: G. Beedle and T. Collins.

—— (1938). *London Revived: Considerations for its Rebuilding in 1666*, ed. E. S. de Beer. Oxford: Clarendon, 1938.

—— (1955). *The Diary of John Evelyn*, ed. E. S. de Beer. 6 vols. Oxford: Clarendon.

Fernie, Eric (1990). 'A Beginner's Guide to the Study of Architectural Proportions and Systems of Length'. In *Medieval Architecture and Its Intellectual Context*, ed. Eric Fernie and Paul Crossley, 229–37. London: Hambledon.

Frangenberg, Thomas (1994). 'Chorographies of Florence: The Use of City Views and City Plans in the Sixteenth Century'. *Imago Mundi*, 46, 41–64.

Friedman, David (2001). '"Fiorenza": Geography and Representation in a Fifteenth-century City View'. *Zeitschrift für Kunstgeschichte*, 64, no. 1, 56–77.

Gent, Lucy (1981). *Picture and Poetry 1560–1620: Relations between Literature and the Visual Arts in the English Renaissance*. Leamington Spa: James Hall.

Geraghty, Anthony (1999). 'Introducing Thomas Laine: Draughtsman to Sir Christopher Wren'. *Architectural History*, 42, 240–45.

—— (2000). 'Nicholas Hawksmoor and the Wren City Church Steeples'. *Georgian Group Journal*, 10, 1–14.

—— (2001). 'Edward Woodroofe: Sir Christopher Wren's First Draughtsman'. *The Burlington Magazine*, 143, no. 1181, 474–79.

—— (2002). 'Wren's Preliminary Design for the Sheldonian Theatre'. *Architectural History*, 45, 275–88.

—— (2004). 'Robert Hooke's Collection of Architectural Books and Prints'. *Architectural History*, 47, 113–25.

—— (2007). *The Architectural Drawings of Sir Christopher Wren at All Souls College, Oxford: A Complete Catalogue*. Aldershot: Lund Humphries.

Gibson, Strickland (1931) (ed.). *Statuta Antiqua Universitatis Oxoniensis*. Oxford: Clarendon.

Gibson, William C. (1970). 'The Biomedical Pursuits of Christopher Wren'. *Medical History*, 14, 331–41.

G[illiflower], M. (1707). *Mercurius Oxoniensis, or the Oxford Intelligencer for 1707*. London: E. Sanger.

Girouard, Mark (1956). 'Three Gothic Drawings in the Smithson Collection'. *Journal of the Royal Institute of British Architects*, 64, no. 1, 35–6.

—— (1962). 'The Smythson Collection of the Royal Institute of British Architects'. *Architectural History*, 5, 23–184.

—— (1983). *Robert Smythson and the Elizabethan Country House*. 2nd edn. New Haven and London: Yale University Press.

Gordon, D. J. (1949). 'Poet and Architect: The Intellectual Setting of the Quarrel between Ben Jonson and Inigo Jones'. *Journal of the Warburg and Courtauld Institutes*, 12, 152–78.

Goudeau, Jeroen (2006–7). 'A Northern Scamozzi: Nicolaus Goldmann and the Universal Theory of Architecture'. *Annali di Architettura*, 18–19, 235–48.

Gough, Richard (1786–96). *Sepulchral Monuments in Great Britain*. 3 vols. London: J. Nichols.

Grew, Nehemiah (1681). *Musaeum Regalis Societatis, or, a catalogue & description of the natural and artificial rarities belonging to the Royal Society and preserved at Gresham Colledge*. London: for the Author.

Gunter, Edmund (1636). *The Description and Use of the Sector*. 2nd edn. London: James Bowler.

—— (1662). *The Works of Edmund Gunter*. 4th edn. London: Francis Eglesfield.

Gunther, R. T. (1937). *Early Science in Cambridge*. Oxford: Oxford University Press.

—— (1923–45). *Early Science in Oxford*. 14 vols. Oxford: Oxford Historical Society [and others].

Hale, J. R. (1982). 'The Defence of the Realm, 1485–1558'. In Colvin (1963–82), iv. 367–401.

—— (1983). 'The Early Development of the Bastion: An Italian Chronology c.1450–c.1534'. In *Renaissance War Studies*, 1–29. London: Hambledon.

Halfpenny, William (1724). *Practical Architecture*. London: T. Bowles, J. Batley, J. Bowles.

—— (1728). *Magnum in Parvo, or, the Marrow of Architecture*. London: J. Wilcox and T. Heath.

—— (1749). *A New and Compleat System of Architecture*. London: John Brindley.

Hambly, Maya (1988). *Drawing Instruments 1580–1980*. London: Sotheby's.

Hamilton, Stanley B. (1933–4). 'The Place of Sir Christopher Wren in the History of Structural Engineering'. *Transactions of the Newcomen Society*, 14, 27–42.

Hanson, T. W. (1928). 'Halifax Builders in Oxford'. *Transactions of the Halifax Antiquarian Society*, 25, 253–317.

Harris, Eileen (1967). 'Burke and Chambers on the Sublime and Beautiful'. In *Essays in the History of Architecture Presented to Rudolf Wittkower on his 65th Birthday*, ed. Douglas Fraser, Howard Hibbard and Milton J. Lewine, 207–13. London: Phaidon.

—— (1990). *British Architectural Books and Writers 1556–1785*. Cambridge: Cambridge University Press.

Harris, John (1970). *Sir William Chambers, Knight of the Polar Star*. London: Zwemmer.

—— (1979). *The Artist and the Country House: A History of Country House and Garden View Painting in Britain, 1540–1870*. London: Sotheby Parke Bernet.

Harris, John, and Higgott, Gordon (1989). *Inigo Jones: Complete Architectural Drawings*. London: Zwemmer.

Harris, John, and Hradsky, Robert (2007). *A Passion for Building: The Amateur Architect in England 1650–1850*. London: Sir John Soane's Museum.

Harris, John, Lever, Jill, and Richardson, Margaret (1983). *Great Drawings from the Collection of the Royal Institute of British Architects*. London: Trefoil.

Harris, John, Orgel, Stephen, and Strong, Roy (1973). *The King's Arcadia: Inigo Jones and the Stuart Court*. London: Arts Council of Great Britain.

Harris, John, and Tait, A. A. (1979). *Catalogue of the Drawings by Inigo Jones, John Webb and Isaac De Caus at Worcester College, Oxford*. Oxford: Clarendon.

Hart, Vaughan (2002). *Nicholas Hawksmoor: Rebuilding Ancient Wonders*. New Haven and London: Yale University Press.

Harvey, John H. (1949). *An Introduction to Tudor Architecture*. London: Art and Technics.

—— (1952). 'Four Fifteenth-century London Plans'. *London Topographical Record*, 20, 1–8.

—— (1953). 'Early Tudor Draughtsmen'. In *The Connoisseur Coronation Book, 1953*, ed. L. G. G. Ramsey, 97–102. London: The Connoisseur.

—— (1972). *The Mediaeval Architect*. London: Wayland.

Harvey, John H., and Oswald, Arthur (1987). *English Mediaeval Architects: A Biographical Dictionary Down to 1550*. 2nd edn. Gloucester: Alan Sutton.

Harvey, P. D. A. (1980). *The History of Topographical Maps: Symbols, Pictures and Surveys*. London: Thames and Hudson.

—— (1981). 'The Portsmouth Map of 1545 and the Introduction of Scale Maps into England'. In *Hampshire Studies*, ed. John Webb, Nigel Yates, and Sarah Peacock, 33–49. Portsmouth: City Record Office.

—— (1993a). 'Estate Surveyors and the Spread of the Scale-map in England 1550–80'. *Landscape History*, 15, 37–49.

—— (1993b). *Maps in Tudor England*. London: PRO and the British Library.

Heyman, Jacques (1996). 'Poleni's Problem'. In *Arches, Vaults and Buttresses: Masonry Structures and Their Engineering*, 345–67. Aldershot: Variorum.

—— (1998a). 'Hooke's Cubico-parabolical Conoid'. *Notes and Records of the Royal Society*, 52, no. 1, 39–50.

—— (1998b). *Structural Analysis: A Historical Approach*. Cambridge: Cambridge University Press.

Higgott, Gordon (1992). 'Varying with Reason: Inigo Jones's Theory of Design'. *Architectural History*, 35, 51–77.

—— (2004a). 'The Fabric to 1670'. In Keene *et al.* (2004), 171–90.

—— (2004b). 'The Revised Design for St Paul's Cathedral, 1685–90: Wren, Hawksmoor and Les Invalides'. *The Burlington Magazine*, 146, no. 1217, 534–47.

Higton, Hester K. (1996). 'Elias Allen and the Role of Instruments in Shaping the Mathematical Culture of Seventeenth-century England'. Unpublished PhD thesis, University of Cambridge.

Hislop, Malcolm (2002). *Medieval Masons*. Princes Risborough: Shire.

Hogarth, William (1753). *The Analysis of Beauty*. London: for the author.

Höltgen, Karl Josef (1990). 'An Unknown Manuscript Translation by John Thorpe of Du Cerceau's *Perspective*'. In *England and the Continental Renaissance*, ed. Edward Chaney and Peter Mack, 215–28. Woodbridge: Boydell.

Hooke, Robert (1665). *Micrographia*. London: J. Martyn and J. Allestry.

—— (1676 [1675]). *A Description of Helioscopes and Some Other Instruments*. London: John Martyn.

—— (1678). *Lectures and Collections*. London: J. Martyn.

—— (1705). *The Posthumous Works*, ed. Richard Waller. London: S. Smith and B. Walford.

Howard, Maurice (2007). *The Building of Elizabethan and Jacobean England*. New Haven and London: Yale University Press.

Hunter, Michael (1995). 'The Making of Christopher Wren'. In *Science and the Shape of Orthodoxy: Intellectual Change in Late Seventeenth-century Britain*, 45–65. Woodbridge: The Boydell Press.

Jackson, T. W. (1885). 'Dr. Wallis' Letter against Mr. Maidwell, 1700'. In *Collectanea*, ed. C. R. L. Fletcher, i. 269–337. Oxford: Clarendon.

Jagger, Cedric (1983). *Royal Clocks: The British Monarchy and its Timekeepers 1300–1900*. London: Hale.

James, M. R. (1924–25). 'An Early Medieval Sketchbook, no. 1916, in the Pepysian Library, Magdalen College, Cambridge'. *The Walpole Society*, 13, 1–17.

Jardine, Lisa (1999). *Ingenious Pursuits: Building the Scientific Revolution*. London: Little, Brown, and Company.

—— (2002). *On a Grander Scale: The Outstanding Career of Sir Christopher Wren*. London: HarperCollins.

Johnston, Stephen (1991). 'Mathematical Practitioners and Instruments in Elizabethan England'. *Annals of Science*, 48, no. 4, 319–44.

—— (1994a). 'The Carpenter's Rule: Instruments, Practitioners and Artisans in 16th-century England'. In *Proceedings of the XIth International Scientific Instrument Symposium, Bologna, September 1991*, ed. G. Dragoni, A. McConnell, and G. L'E. Turner, 39–45. Bologna: Grafis Edizione.

Bibliography

—— (1994b). 'Making Mathematical Practice: Gentlemen, Practitioners and Artisans in Elizabethan England'. Unpublished PhD thesis, University of Cambridge.

—— (1996). 'The Identity of the Mathematical Practitioner in 16th-century England'. In *Der 'Mathematicus': zur Entwicklung und Bedeutung einer neuen Berufsgruppe in der Zeit Gerhard Mercators*, ed. Irmgarde Hantsche, 93–120. Bochum: Brockmeyer.

—— (2006). 'Reading Rules: Artefactual Evidence for Mathematics and Craft in Early-Modern England'. In Taub and Willmoth (2006), 233–53.

Keene, Derek, Burns, Arthur, and Saint, Andrew (2004) (eds). *St Paul's: The Cathedral Church of London, 604–2004*. New Haven and London: Yale University Press.

Kemp, Martin (1986). 'Geometrical Bodies as Exemplary Forms in Renaissance Space'. In *World Art: Themes of Unity in Diversity: Acts of the XVIth International Congress of the History of Art*, ed. Irving Lavin, 237–42. University Park, PA: Pennsylvania State University Press.

Kirby, Joshua (1754). *Dr. Brook Taylor's Method of Perspective Made Easy*. Ipswich: for the author.

—— (1761). *The Perspective of Architecture*. London: for the author.

Knight, Richard (1988). 'The Carpenter's Rule'. *Newsletter of the Tools and Trades Society*, 20, Winter, 12–19.

—— (1990). 'A Carpenter's Rule from the Mary Rose'. *Tools and Trades*, 6, 43–55.

Knoop, Douglas, and Jones, G. P. (1935). *The London Mason in the Seventeenth Century*. Manchester: Manchester University Press and Quatuor Coronati Lodge.

—— (1936). 'The Rise of the Mason Contractor'. *Journal of the Royal Institute of British Architects*, 43, no. 20, 1061–71.

Knoop, Douglas, Jones, G. P., and Hamer, Douglas (1938). *The Two Earliest Masonic MSS*. Manchester: Manchester University Press.

Lang, Jane (1956). *Rebuilding St. Paul's after the Great Fire of London*. London: Oxford University Press.

Lefèvre, Wolfgang (2004a). 'The Emergence of Combined Orthographic Projections'. In Lefèvre (2004b), 209–44.

Lefèvre, Wolfgang (2004b) (ed.). *Picturing Machines 1400–1700*. Cambridge, MA: MIT Press.

Lever, Jill, and Richardson, Margaret (1984). *The Art of the Architect: Treasures from the RIBA's Collections*. London: Trefoil.

Leybourn, William (1667). *The Line of Proportion or Numbers, Commonly Called Gunters Line*. London: G. Sawbridge.

Lindley, Phillip (1988). 'The Sculptural Programme of Bishop Fox's Chantry Chapel'. *Winchester Cathedral Record*, 57, 33–7. Reprinted in Phillip Lindley, *Gothic to Renaissance: Essays on Sculpture in England* (Stamford: Paul Watkins, 1995), 207–12.

—— (2003). '"The Singuler Mediacions and Praiers of Al the Holie Companie of Heven": Sculptural Functions and Forms in Henry VII's Chapel'. In *Westminster Abbey: The Lady Chapel of Henry VII*, ed. Tim Tatton-Brown and Richard Mortimer, 259–93. Woodbridge: Boydell Press.

Lorini, Buonaiuto (1609). *Le fortificationi*. Venice: F. Rampazetto.

Lotz, Wolfgang (1977). 'The Rendering of the Interior in Architectural Drawings of the Renaissance'. In *Studies in Italian Renaissance Architecture*, 1–65. Cambridge, MA and London: MIT Press.

Louw, Hentie (2006). 'The "Mechanick Artist" in Late Seventeenth-century English and French Architecture'. In *Robert Hooke: Tercentennial Studies*, ed. Michael Cooper and Michael Hunter, 181–99. Aldershot: Ashgate.

Lydgate, John (1934). *The Minor Poems of John Lydgate*, ed. Henry Noble MacCracken. 2 vols. London: Oxford University Press.

Lyle, David (1762). *The Art of Short Hand Improved*. London: A. Millar.

Lyles, Anne, and Hamlyn, Robin (1997). *British Watercolours from the Oppé Collection with a Selection of Drawings and Oil Sketches*. London: Tate Gallery.

Macartney, Mervyn (1907). 'The Present Condition of St. Paul's Cathedral'. *Journal of the Royal Institute of British Architects*, 15, 53–71.

Macdonald, A. (1937). 'Plans of Dover in the Sixteenth Century'. *Archeologia Cantiana*, 49, 108–26.

Macray, William Dunn (1890). *Annals of the Bodleian Library Oxford*. 2nd edn. Oxford: Clarendon.

Malie, Thomas (1737). *A New and Accurate Method of Delineating All the Parts of the Different Orders in Architecture, by Means of a Well Contriv'd, and Most Easily Manag'd Instrument*. London: T. Heath.

Mallgrave, Harry Francis (2005). *Modern Architectural Theory: A Historical Survey, 1673–1968*. Cambridge: Cambridge University Press.

Malton, Thomas (1783). *An Appendix or Second Part to the Compleat Treatise on Perspective*. London: for the author.

Mandosio, Jean Marc (2003). 'Des "mathématiques vulgaires" à la "monade hiéroglyphique": les Eléments d'Euclide vus par John Dee'. *Revue d'Histoire des Sciences*, 56, 475–91.

Marks, Richard, and Williamson, Paul (2003) (eds.). *Gothic: Art for England 1400–1547*. London: V&A Publications.

Marr, Alexander (2004). '"Curious and Useful Buildings": The Mathematical Model of Sir Clement Edmondes'. *Bodleian Library Record*, 18, no. 2, 108–50.

Martyn, John (1661). *Mensuration Made Easie*. London: Thomas Martyn.

Maslen, Keith (1972). 'Printing for the Author: From the Bowyer Printing Ledgers, 1710–1775'. *The Library*, ser. 5, 27, 302–09.

McDermott, James (1998). 'Humphrey Cole and the Frobisher Voyages'. In Ackermann (1998), 15–19.

Merriman, Marcus (1983). 'Italian Military Engineers in Britain in the 1540s'. In *English Map-making 1500–1650*, ed. Sarah Tyacke, 57–67. London: British Library.

Millburn, John R. (2000). *Adams of Fleet Street: Instrument Makers to King George III*. Aldershot: Ashgate.

Milman, Lena (1908). *Sir Christopher Wren*. London: Duckworth.

Minet, William (1922). 'Some Unpublished Plans of Dover Harbour'. *Archaeologia*, 62, 185–225.

Monconys, Balthasar de (1887). *Les Voyages*, ed. Charles Henry. Paris: Hermann. (Reprint of the original edition published in Lyon, 1665–66.)

Moore, Jonas (1673). *Modern Fortification*. London: N. Brooke.

More, Richard (1602). *The Carpenter's Rule*. London: Felix Kyngston.

Morrison-Low, A. D. (1997). '"Spirit of Place": Some Geographical Implications of the English Provincial Instrument Trade, 1760–1850'. *Bulletin of the Scientific Instrument Society*, 53, 19–24.

—— (2007). *Making Scientific Instruments in the Industrial Revolution*. Aldershot: Ashgate.

Morton, Alan Q., and Wess, Jane A. (1993). *Public and Private Science: The King George III Collection*. Oxford: Oxford University Press.

Moxon, Joseph (1679). *Mathematicks Made Easy, or, a Mathematical Dictionary*. London: J. Moxon.

Murdoch, John (1971). 'Euclid: Transmission of the Elements'. In *Dictionary of Scientific Biography*, ed. Charles Coulston Gillispie, iv.437–59. New York: Scribner.

Myres, J. N. L. (1967). 'Recent Discoveries in the Bodleian Library'. *Archaeologia*, 101, 151–68.

Nasmith, James (1778). *Itineraria Symonis Simeonis et Willelmi de Worcestre*. Cambridge: J. Archdeacon.

Newman, John (1988). 'Italian Treatises in Use: The Significance of Inigo Jones's Annotations'. In *Les Traités de la Renaissance*, ed. Jean Guillaume, 435–41. Paris: Picard.

—— (1992). 'Inigo Jones's Architectural Education before 1614'. *Architectural History*, 35, 18–50.

—— (1997). 'The Architectural Setting'. In *The History of the University of Oxford, Volume IV: Seventeenth-century Oxford*, ed. Nicholas Tyacke, 135–77. Oxford: Clarendon.

North, Roger (1981). *Of Building: Roger North's Writings on Architecture*, ed. Howard Colvin and John Newman. Oxford: Clarendon Press.

—— (2000). *Notes of Me: The Autobiography of Roger North*, ed. Peter Millard. Toronto: University of Toronto Press.

Nuti, Lucia (1994). 'The Perspective Plan in the Sixteenth Century: The Invention of a Representational Language'. *Art Bulletin*, 76, no. 1, 105–28.

Oppé, A. P. (1950). *English Drawings, Stuart and Georgian Periods, in the Collection of His Majesty the King at Windsor Castle*. London: Phaidon.

Orgel, Stephen, and Strong, Roy (1973). *Inigo Jones: The Theatre of the Stuart Court*. 2 vols. London: Sotheby Parke Bernet; Berkeley: University of California Press.

Otto, Christian F. (1979). *Space into Light: The Churches of Balthasar Neumann*. New York and Cambridge, MA: Architectural History Foundation and MIT Press.

Oughtred, William (1652). *Clavis Mathematicae*. Oxford: L. Lichfield.

Owen, Felicity (1995–96). 'Joshua Kirby (1716–74): A Biographical Sketch'. *Gainsborough's House Review*, 61–76.

Pacey, Arnold (2007). *Medieval Architectural Drawing*. Stroud: Tempus.

Palladio, Andrea (1601). *I quattro libri dell'architettura*. Venice: B. Carampello.

—— (1663). *The First Book of Architecture*, tr. Godfrey Richards. London: G. Richards.

Paulson, Ronald (1989). *Hogarth's Graphic Works*. 3rd edn. London: Print Room.

—— (1991–93). *Hogarth*. 3 vols. New Brunswick, NJ: Rutgers University Press.

Peach, C. Stanley, and Allen, W. Godfrey (1930). 'The Preservation of St Paul's Cathedral'. *Journal of the Royal Institute of British Architects*, 37, no. 18, 656–76.

Peacham, Henry (1612). *Graphice, or the Most Auncien and Excellent Art of Drawing and Limning Disposed into Three Bookes*. 2nd edn. London: W.S. for John Browne.

Pepys, Samuel (1926). *Private Correspondence and Miscellaneous Papers: 1679–1703*, ed. J. R. Tanner. 2 vols. London: Bell.

Perkins, Jocelyn (1938–52). *Westminster Abbey: Its Worship and Ornaments*. 3 vols. London: H. Milford.

Philip, Ian (1983). *The Bodleian Library in the Seventeenth and Eighteenth Centuries*. Oxford: Clarendon.

Plot, Robert (1677). *The Natural History of Oxford-shire*. Oxford: the Theater.

Poleni, Giovanni (1748). *Memorie istoriche della gran cupola del Tempio Vaticano*. Padova.

Poley, Arthur F. E. (1927). *St. Paul's Cathedral, London: Measured, Drawn & Described*. London: for the author.

Poole, Rachel (1912–25). *Catalogue of Portraits in the Possession of the University, Colleges, City, and County of Oxford*. 3 vols. Oxford: Clarendon.

—— (1922). 'The Architect of the Schools and the Tower of the Five Orders'. *Bodleian Quarterly Record*, 3, no. 35, 263–64.

Popplow, Marcus (2004). 'Why Draw Pictures of Machines? The Social Context of Early Modern Machine Drawings'. In Lefèvre (2004b), 17–48.

Pratt, Roger (1928). *The Architecture of Roger Pratt*, ed. R. T. Gunther. Oxford: University of Oxford Press.

Price, Derek J. (1955). 'Medieval Land Surveying and Topographical Maps'. *The Geographical Journal*, 121, no. 1, 1–7.

Rambaldi, Enrico I (1989). 'John Dee and Federico Commandino: An English and an Italian Interpretation of Euclid During the Renaissance'. In *Italy and the English Renaissance*, ed. Sergio Rossi and Dianella Savoia, 123–53. Milan: Unicopli.

Ramsay, Allan (1755). *The Investigator. Number CCCXXII. To Be Continued Occasionally*. London: A. Millar.

Recorde, Robert (1551). *The Pathway to Knowledg, Containing the First Principles of Geometrie*. London: R. Wolfe.

Reddaway, T. F. (1951). *The Rebuilding of London after the Great Fire*. London: Arnold.

Renn, Jürgen, and Valleriani, Matteo (2001). 'Galileo and the Challenge of the Arsenal'. *Nuncius*, 16, no. 2, 481–503.

Roberts, Jane (1987). *Royal Artists: From Mary Queen of Scots to the Present Day*. London: Grafton.

—— (1997). 'Sir William Chambers and Geoge III'. In *Sir William Chambers: Architect to George III*, ed. John Harris and Michael Snodin, 41–54. New Haven and London: Yale University Press.

—— (2004) (ed.). *George III and Queen Charlotte: Patronage, Collecting, and Court Taste*. London: Royal Collection.

Robertson, John (1747). *Treatise of Such Mathematical Instruments, as are Usually Put into a Portable Case*. London: T. Heath.

Robinson, H. W., and Adams, W. (1935). *The Diary of Robert Hooke, 1662–1680*. London: Taylor and Francis.

Robinson, William (1736). *Proportional Architecture*. 2nd edn. London: W. Dicey.

Rouquet, Jean André (1755). *The Present State of the Arts in England*. London: J. Nourse.

Royal Institute of British Architects (1961). *Architectural Drawings from the Collection of the Royal Institute of British Architects*. London: A. Tiranti.

—— (1972). *Great Drawings from the Collection*. London: RIBA.

Salzman, L. F. (1952). *Building in England down to 1540: A Documentary History*. Oxford: Clarendon Press.

Saunders, Andrew (1989). *Fortress Britain: Artillery Fortification in the British Isles and Ireland*. Liphook: Beaufort.

Sayce, R. A. (1970). 'Foreword'. In Allsopp (1970).

Scamozzi, Vincenzo (1615). *L'idea della architettura universale*. Venice: expensis auctoris.

—— (1669). *The Mirrour of Architecture*. London: William Fisher.

Schulz, Juergen (1978). 'Jacopo de' Barbari's View of Venice: Map Making, City Views, and Moralized Geography before the Year 1500'. *Art Bulletin*, 60, no. 3, 425–74.

Sekler, Eduard F. (1956). *Wren and His Place in European Architecture*. London: Faber and Faber.

Serlio, Sebastiano (1611). *The First Booke of Architecture … Entreating of Geometrie*. London: R. Peake.

—— (1996–2001). *On Architecture*, ed. Vaughan Hart and Peter Hicks. 2 vols. New Haven and London: Yale University Press.

Shelby, Lon R. (1961). 'Medieval Masons' Tools: The Level and the Plumb Rule'. *Technology and Culture*, 2, 127–30.

—— (1964). 'The Role of the Master Mason in Mediaeval English Building'. *Speculum*, 39, no. 3, 387–403.

—— (1965). 'Medieval Masons' Tools II: Compass and Square'. *Technology and Culture*, 6, 236–48.

—— (1967). *John Rogers: Tudor Military Engineer*. Oxford: Oxford University Press.

—— (1972). 'The Geometrical Knowledge of Medieval Master Masons'. *Speculum*, 47, no. 3, 395–421.

Shelby, Lon R. (1977) (ed.). *Gothic Design Techniques: The Fifteenth-century Design Booklets of Mathes Roriczer and Hanns Schmuttermayer*. Carbondale, IL: Southern Illinois University Press.

Shirley, John W. (1983). *Thomas Harriot: A Biography*. Oxford: Clarendon Press.

Shute, John (1563). *The First and Chief Groundes of Architecture*. London: Thomas Marshe.

Simpkins, Diana M. (1966). 'Early Editions of Euclid in England'. *Annals of Science*, 22, no. 4, 225–49.

Skelton, R. A. (1970). 'The Military Surveyor's Contribution to British Cartography in the 16th Century'. *Imago Mundi*, 24, 77–83.

Skelton, R. A., and Harvey, P. D. A. (1986) (eds.). *Local Maps and Plans from Medieval England*. Oxford: Clarendon.

Skempton, A. W., and Chrimes, M. M. (2002) (eds.). *Biographical Dictionary of Civil Engineers in Great Britain and Ireland*. London: Thomas Telford.

Sladen, Teresa (2004). 'Embellishment and Decoration, 1696–1900'. In Keene *et al.* (2004), 233–56.

Smart, Alastair (1992). *Allan Ramsay: Painter, Essayist and Man of the Enlightenment*. New Haven and London: Yale University Press.

Smith, Angela (1988). 'The Chantry Chapel of Bishop Fox'. *Winchester Cathedral Record*, 57, 27–32.

Soo, Lydia M. (1998). *Wren's 'Tracts' on Architecture and Other Writings*. Cambridge: Cambridge University Press.

Stancliffe, Martin (2004). 'The Conservation of the Fabric'. In Keene *et al.* (2004), 293–303.

Stephens, Frederic George, and George, Mary Dorothy (1870–1954). *Catalogue of Political and Personal Satires in the Department of Prints and Drawings in the British Museum*. 11 vols. London: British Museum.

Stillwell, John (2004). *Mathematics and its History*. New York: Springer.

Bibliography

Stuart, James, and Revett, Nicholas (1762–1830). *The Antiquities of Athens: Measured and Delineated*. 5 vols. London.

Summerson, John (1953). *Sir Christopher Wren*. London: Collins.

—— (1957–58). 'Three Elizabethan Architects'. *Bulletin of the John Rylands Library*, 40, 202–28.

—— (1959). 'The Building of Theobalds, 1564–1585'. *Archaeologia*, 97, no. 47, 107–26.

—— (1964–66). 'The Book of Architecture of John Thorpe in Sir John Soane's Museum'. *Walpole Society*, 40, complete issue.

—— (1975). 'The Works from 1547–1660'. In Colvin (1963–82), vol. 3.

—— (1990a). 'Christopher Wren: Why Architecture?' In *The Unromantic Castle and Other Essays*, 63–68. London: Thames and Hudson.

—— (1990b). 'J. H. Mansart, Sir Christopher Wren and the Dome of St Paul's Cathedral'. *The Burlington Magazine*, 132, no. 1042, 32–6.

—— (1990c). 'John Thorpe and the Thorpes of Kingscliffe'. In *The Unromantic Castle and Other Essays*, 17–40. London: Thames and Hudson.

Tait, A. A. (1978). 'Inigo Jones's "Stone-Heng"'. *The Burlington Magazine*, 120, no. 900, 155–59.

—— (1987). 'Post-modernism in the 1650s'. In *Inigo Jones and the Spread of Classicism: Georgian Group Symposium 1986*, ed. John Newman, 23–35. London: The Georgian Group.

Taub, Liba, and Willmoth, Frances (2006) (eds). *The Whipple Museum of the History of Science: Instruments and Interpretations*. Cambridge: Whipple Museum.

Taylor, E. G. R. (1947). 'The Surveyor'. *The Economic History Review*, 17, no. 2, 121–33.

—— (1954). *The Mathematical Practitioners of Tudor and Stuart England*. Cambridge: Cambridge University Press.

Tesseract (2005). *Early Scientific Instruments: Catalogue 81*. Hastings-on-Hudson, NY: Tesseract.

Toker, Franklin (1985). 'Gothic Architecture by Remote Control: An Illustrated Building Contract of 1340'. *Art Bulletin*, 67, no. 1, 67–95.

Truesdell, Clifford A. (1960). *The Rational Mechanics of Flexible or Elastic Bodies: 1638–1788*, Leonhardi Euleri Opera Omnia, Series Secunda, Volumina XI, Sectio Secunda. Zurich: Orell Füssli.

Turnbull, G. H. (1953). 'Samuel Hartlib's Influence on the Early History of the Royal Society'. *Notes and Records of the Royal Society*, 10, no. 2, 101–30.

Turner, A. J. (1974). 'Mathematical Instruments and the Education of Gentlemen'. *Annals of Science*, 30, 51–88.

—— (1987). *Early Scientific Instruments, Europe 1400–1800*. London: Sotheby's.

—— (1994). 'Natural Philosophers, Mathematical Practitioners and Timber in Later 17th Century England'. *Nuncius*, 9, 619–34.

Turner, G. L'E. (1967). 'The Auction Sales of the Earl of Bute's Instruments, 1793'. *Annals of Science*, 23, 213–242.

—— (1992). 'Queen Charlotte's Protractor and Joshua Kirby'. *Bulletin of the Scientific Instrument Society*, 35, 23–4.

—— (2000). *Elizabethan Instrument Makers: The Origins of the London Trade in Precision Instrument Making*. Oxford: Oxford University Press.

Tuttle, Richard J., Adorni, Bruno, Frommel, Christoph Luitpold, and Thoenes, Christof (2002). *Jacopo Barozzi da Vignola*. Milan: Electa.

Uffenbach, Zacharias Konrad von (1928). *Oxford in 1710: From the Travels of Zacharias Conrad Von Uffenbach*, ed. W. H. Quarrell and W. J. C. Quarrell. Oxford: Blackwell.

Uglow, Jenny (1997). *Hogarth*. London: Faber and Faber.

Vignola, Giacomo Barozzi da (1607). *Regola delli cinque ordini d'architettura*. Rome: Andreas Vaccarius.

—— (1999). *Canon of the Five Orders of Architecture*, ed. Branko Mitrovic. New York: Acanthus.

Ward, F. A. B. (1981). *A Catalogue of European Scientific Instruments in the Department of Medieval and Later Antiquities of the British Museum*. London.

Ward, John (1740). *The Lives of the Professors of Gresham College*. London: J. Moore for the author.

Ware, Samuel (1809). *A Treatise of the Properties of Arches, and Their Abutment Piers*. London: J. Taylor.

Warren, Edward Prioleau (1923). 'Sir Christopher Wren's Repair of the Divinity School and Duke Humphrey's Library, Oxford'. In Dircks (1923), 233–38.

Watkin, David (2004). *The Architect King: George III and the Culture of the Enlightenment*. London: Royal Collection.

Webb, John (1725). *A Vindication of Stone-Heng Restored*. 2nd edn. London: G. Conyers. (First edn 1665.)

Whinney, Margaret (1971). *Wren*. London: Thames and Hudson.

Wilkins, John (1648). *Mathematicall Magick, or, the wonders that may be performed by mechanicall geometry*. London: Gellibrand.

Willis, Robert, and Clark, John Willis (1988). *The Architectural History of the University of Cambridge*. 4 vols. Cambridge: Cambridge University Press. (First edn 1886.)

Willis, Thomas (1664). *Cerebri Anatome cui Accessit Nervorum Descriptio et Usus*. London: J. Martyn and J. Allestry.

Willmoth, Frances (1993). *Sir Jonas Moore: Practical Mathematics and Restoration Science*. Woodbridge: Boydell.

Wilton-Ely, John (1977). 'The Rise of the Professional Architect in England'. In *The Architect: Chapters in the History of the Profession*, ed. Spiro Kostof, 180–208. Oxford: Oxford University Press.

Wittkower, Rudolf (1974a). 'Classical Theory and Eighteenth-century Sensibility'. In Wittkower (1974d), 193–204.

—— (1974b). 'English Literature on Architecture'. In Wittkower (1974d), 95–112.

—— (1974c). 'Inigo Jones, Architect and Man of Letters'. In Wittkower (1974d), 51–64.

—— (1974d). *Palladio and English Palladianism*. London: Thames and Hudson.

Woodman, Francis (1992). 'The Waterworks Drawings of the Eadwine Psalter'. In *The Eadwine Psalter: Text, Image, and Monastic Culture in Twelfth-Century Canterbury*, ed. Margaret Gibson, T. A. Heslop, and Richard W. Pfaff, 168–77. London and University Park, PA: Modern Humanities Research Association and Pennsylvania State University Press.

Worcestre, William (1969). *Itineraries*, ed. John Harvey. Oxford: Clarendon Press.

—— (2000). *The Topography of Medieval Bristol*, ed. Frances Neale. Bristol: Bristol Record Society.

Worsley, Giles (1991). *Architectural Drawings of the Regency Period, 1790–1837*. London: André Deutsch.

—— (1994). 'The Gentleman-Professional'. In *The Role of the Amateur Architect: Georgian Group Symposium 1993*, ed. Giles Worsley, 14–20. London: The Georgian Group.

Worsop, Edward (1582). *A Discouerie of Sundrie Errours and Faults Daily Committed by Lande-Meaters, Ignorant of Arithmetike and Geometrie*. London: Gregorie Seton.

Wotton, Henry (1624). *The Elements of Architecture*. London: John Bill.

Wren, Christopher, Jr (1965). *Parentalia*. Farnborough: Gregg Press (reproduction of the 1750 edition of the 'Heirloom' copy at RIBA).

Wu, Nancy Y. (2002) (ed.). *Ad Quadratum: The Practical Application of Geometry in Medieval Architecture*. Aldershot: Ashgate.

Yates, Frances A. (1969). *Theatre of the World*. London: Routledge & Kegan Paul.

Yeomans, David (1992). *The Trussed Roof: Its History and Development*. Aldershot: Scolar.

—— (1997). 'The Serlio Floor and Its Derivations'. *Architectural Research Quarterly*, 2, no. 3, 74–83.

Index

Page numbers in italics refer to figures; numbers preceded by 'cat' refer to catalogue numbers

Adams, George 133, *138–41*, 141–2, *143*, cat70–75
Adams, Robert 55
Archisesto per formar con facilità li cinque ordini d'Architettura (Revesi Bruti) 124
architectonic drawing board 122–4, *124*, cat56
architectonic rule *129*, 129
architectonic sector 116, *117*, 120, 121–2, *127*, 136–7, *138–9*, 141, cat52, cat55, cat59, cat70–71
architectonic sliding plates 122, *122*, cat56
architectural drawing
    medieval 18–19, *19*, 21–30, *21–5*, *27*, cat1, cat4–6
    16th century 28–30, *29–30*, *58*, 60, cat7–9
    17th century 67–71, *68–70*, 77–8, 80, cat29–30, cat35
    George III *132–7*, 133–6, cat62–68
architectural instruments
    17th century 111–17, *112–13*, *116*, cat52
    18th century 120–9, *120*, *122*, *125*, *127*, cat55–56, cat58–59, cat70–71
architectural model (Edmondes) 72–5, *73*, 76, cat32
architectural practice 11
    medieval 17
    16th century 31–3
    17th century 65, 83
architectural protractors *125*, cat58
Art of Shorthand Improved (Lyle) 142
astronomical instruments 51–3, *52*, cat18
axonometric drawing 98

Blocks for Hogarth's Wigs (Sandby) 148–9, *149*, cat77
Blum (or Bloome), Hans 80, *80*, cat36
Bodleian tower *74*, 75–6
Boke Named Tectonicon, A (Digges) 46–9, *46–8*, cat15–16
book collections 79
Booke of Five Collumnes of Architecture, The (Blum) 80–1, *80*, cat36
Boulogne, topographical survey *40*, 41
Braun, Georg 35
Bristol, St Stephen's Church *21*, 22, cat4
Brown, John 111–12, *112*
builder's level, 19th century *20*, cat3
Burke, Edmund 131
Bute, John Stuart, Earl of 133, 141–2, 146, 148

Calais
    Guines Castle *36*, 39, 45
    town and harbour *34*, 35, cat10
Cambridge
    Hospital of St John the Evangelist 19, *19*, cat1
    Trinity Library 98, *98*
Canterbury Cathedral *24*, 26
carpenter's rule *48*, 50–4, *53–4*, *55*, cat19–20
Carpenter's Rule, The (More) 46
Carwitham, Thomas 120–2, *121–2*, cat55–56
Cerebri Anatome (Willis) 89, 91
Chambers, William 133, 151
Charterhouse, London *25*, 26
chorography 33, 35, 41–2
Civitates orbis terrarum (Braun and Hogenberg) 35
Cole, Humfrey 50–1, *50–1*, 55, 62, cat17, cat27
compass, 18th century *20*, cat2
Cooke Manuscript 17, 18
Cosmographical Glasse, The (Cuningham) *32*, 35
Cuningham, William *32*, 33, 35
Cursitors' Hall 57, 59–60, cat23

Description and Use of an Architectonick Sector, The (Carwitham) 121
Description and Use of the Sector, The (Gunter) *112–13*
Dickinson, William 102, 104, 167–9, cat47
Digges, Leonard 45–51, *46–8*, cat15–16
Dover harbour *33*, 35, 56, *56*, 59, cat22
Dr Brook Taylor's Method of Perspective Made Easy (Kirby) *144*, 145–6
draughtsmen 100
drawing *see* architectural drawing; plan drawings
drawing board, architectonic 122–4, *124*, cat56
drawing instruments
    16th century 60, *61*, cat26
    17th century *115*, 117, cat51
    18th century *119*, 120, *128*, 128, *130*, 130, cat54, cat61
    see also architectural instruments
Du Cerceau, Jacques Androuet 81–2, *81*, cat37
Duke Humphrey's Library, Oxford 98–9, *99*

Edmondes, Clement 72–6, *76*, cat32
Elements of Geometrie (Euclid) 17, cat34
engraved maps *32*, 35
estate plans 41–2, *42–3*, cat13–14
Euclidian geometry 75, cat34

and the 'art of masonry' 17–18, *18*, 19
Evelyn, John 83, 84, 163

First Booke of Architecture, The (Serlio) 123
Fisher, William 111, 114
Five Orders of PERRIWIGS . . ., The (Hogarth) 146, *147*, cat76
folding rule 50–1, 53–4, *54–5*, cat17, cat21
fortification sector *116*, 117, cat52
Fortificationi, Le (Lorini) 71, *72*, cat31
fortifications *see* military architecture

Gemini, Thomas 49, *51*, *52*, cat18
geometrical methods 67–72, 75, 98, 100–2, 162–3, cat34, cat45
geometrical model (Edmondes) 72–5, *73*, 76, cat32
geometrical square 60, 62
geometrical surveys 39–41
geometry, and the 'art of masonry' 17–18, *18*, 19
George III
    architectural drawings *132–7*, 133–6, cat62–68
    architectural instruments 137, 141
    artistic patronage 136
Gregory, David 163
Grignion, C. 131
Guines Castle *36*, 39, 45
Gunter, Edmund 112–14, *112–13*

Halfpenny, William 122–4, *123*, *124–5*, 129, *129*, cat57–58
Hardouin-Mansart, Jules 157, 159
Hawksmoor, Nicholas 100, 102, *102*, 107–9, *109*, 152–3, 155, 161, *161*, 164, *164*, cat46, cat48–50
Heath, Thomas 120, *121*, 122, *122*, 124, 126, *126*, 129–30, *130*, 142, cat55–56, cat58–59, cat61
Henry VI tomb *28*, 29, cat7
Hogarth, William 131, 144–5, *145–9*, *147–8*, cat76–77
Hogenberg, Franz 35
Hooke, Robert 92, 97, 101, 155, 163–4
Hull 41
    defensive works *37*, 39–40, cat11
    Manor 42, *42–3*, cat13–14

Idea della architettura universale (Scamozzi) 111
instrument makers 142
    16th century 50–5, 62
    17th century 111–1

Index

Instrument makers continued
    18th century 120–30, 133, 141–2
    see also architectural instruments; drawing instruments; mathematical instruments; surveying instruments
Invalides, Les 157, *158*, 159, *160*
Isidore of Seville 17–18
Itineraries (William Worcestre) 21–2

'joint rule' 111–12, *112*
Jones, Inigo 65–72, *68–70*, 83, cat29–31

Kip, Jan 166
Kirby, Joshua 136–42, *137–40*, 144–5, 145–6, cat69–72

land surveys 33, 35
Leçons de perspective positive (Du Cerceau) 81–2, *81*, cat37
Lee, Richard *34*, 35, *36*, 39, 41, 44, cat10
Lepautre, Pierre 159, *160*
level, 19th century *20*
Leybourn, William 111
logarithmic line 53, 55, cat21
London
    Charterhouse *25*, 26
    Middle Temple 83–4
    old St Pauls *93*, 94–5, *105*, cat43
    plot and tenements (*c*.1475) 26, *27*, cat6
    Royal Society 84
    St Paul's Cathedral *see* St Paul's Cathedral
    Westminster Abbey *28*, 29
    Wren's rebuilding plans 95–6, *95*, cat44
Lorini, Buonaiuto *71*, 72, cat31
Lydgate, John 18, 21
Lyle, David *140–1*, 142, cat73–74

Magdalene College, Cambridge 22, *22*
Magnum in Parvo (Halfpenny) 122, *123*, *124*, cat57
Malie, Thomas 124, *126*, *126*
Malton, Thomas 142, 145
Marot, Jean *158*
Marrow of Architecture, The (Halfpenny) 122, *123*, *124*, cat57
Mary Rose (ship) 48, *48*, 50
masonry, and the 'science of geometry' 17–18, 19
masons' drawings 18–19, *19*, 21, *58*, 60, cat1
masons' templates 18–19, 21, 22
masons' tools *20*, 21, cat2–3
mathematical instruments 111–14, *112–13*, 115, 117, 121, 126, 128, cat51
    see also architectural instruments
mathematical model (Edmondes) 72–5, *73*, 76, cat32
mathematical playing cards *118*, 119–20, cat53
mathematical practice 12–13, 45, 49
measured surveys 35, *36*, *38–40*, 39–44, *42–3*, cat13–14
measuring rulers 48, 49, 50, *50–1*, cat17, cat19–21
medieval drawing 18–19, *19*, 21–30, *21–5*, *27*, cat1, cat4–6
medieval masonry, and the 'science of geometry' 17–18
medieval plans 22–9, *23–5*, *27*, *30*, *30*

medieval surveys 26, *27*, 29, cat6
medieval tools *20*, 21, cat2–3
microscope (for George III) *143*, cat75
Middle Temple 83–4
military architecture 31–3, 39–41, 72, cat10–12
Mirrour of Architecture, The (Scamozzi) 111, *111–12*, 114
modular design methods 67, 80
Moore, Jonas *116*, cat52
More, Richard 46
Moxon, Joseph 119

New and Accurate Method, A (Malie) 126, *126*
New and Compleat System of Architecture, A (Halfpenny) 129
Newnham, Hampshire *33*, 35
Newsam, Bartholomew *60*, cat26
North, Roger 83–4, 114–19
Norwich *32*, 35

Office of Works (Royal) 55, 61, 63
Office of Works (St Paul's) 104–5
Oughtred, William 84, 85
Oxford
    Duke Humphrey's Library 98–9, *99*
    Sheldonian Theatre *92*, 94, 97, 98, *98*, cat42

Palladio, Andrea 67, *70*, 71, *77*, 80
Pantometria (Digges) *60*, 62
Parentalia (Wren) 85, *86*, 98
Peake, Robert 54–5
perspective, representation of 80–1, 85, 136, 142, 144, 146
Perspective of Architecture (Kirby) 136, *137*, 145, 146, cat69
Pett, Phineas 63, *63*
pictorial conventions 26, 35, 41–2
plan drawings
    medieval 22–9, *23–5*, *27*, cat5
    16th century 32–44, *32–4*, *36–40*, *42–3*, *56*, cat10–12, cat22–23
plattes/plats 22, 26, 32, 39
playing cards, mathematical *118*, 119–20, cat53
Poleni, Giovanni 155, *156*
Poley, Arthur *154*, 156, *168*, *169*
Portington, William 63, *63*
Portsmouth, survey plan *38*, *39*, 40–1, cat12
Practical Architecture (Halfpenny) 124
Pratt, Roger 83
Present State of the Arts in England, The (Jean André Rouquet) 149
Primo Libro (Serlio) 17
printed maps *32*, 35
proportional design methods 67, *68–9*, 71, 80
protractors *125*, *140*, 142, cat58, cat72

Quadratum Geometricum 62
Quattro libri dell'architettura, I (Palladio) 67, 69–71, *70*, *77*, cat30
Quinque columnarum exacta descriptio atque deliniatio (Blum) 80–1, *80*

Ramsay, Allan 131–2
Recorde, Robert 18

Regius Manuscript 17, 18
Regola delli cinque ordini (Vignola) 67, *68*, cat29
rescaling device 122–4, *123–4*, cat57
    see also architectonic sector
Revesi Bruti, Ottavio 124, *126*, cat59
Robertson, John *128*, *128*, 130, cat60
Rogers, John *36*, *37*, 39–40, *40*, 41, *42*, *42–3*, 44, cat11, cat13–14
Rouquet, Jean André 149
Royal Academy of Arts 131
Royal Mathematical School 114
Royal Society 84
Ryther, Augustine 55

Sandby, Paul *149*, cat77
Saunders, Samuel *119*, 120, cat54
Savile, Henry 74, *74*, 76, cat33
scale plans and maps *36*, 39–44
Scamozzi, Vincenzo 111, *111–12*
Scarburgh, Charles 85
sector 111–14, *112–13*
    see also architectonic sector
Serlio, Sebastiano 17, 18, 54, 80, 123, *123*
Sheldonian Theatre, Oxford *92*, 94, 97, 98, *98*, cat42
Shute, John 54
Simons, Ralph 63–4, *64*, cat28
Smythson, Robert *58*, 60, cat24
Society of Civil Engineers 131
Specchi, Alessandro 159
St John the Evangelist, Cambridge 19, *19*, cat1
St Paul's Cathedral 69, *69*, 97, 99–100, *101–3*, 106–9, cat47–48
    dome design *101*, 102–4, 152–4, 155–69, *158*, *159*, *160–1*, *165*, cat45–47, cat49–50
    old St Pauls *93*, 94–5, *105*, cat43
St Peter's, Rome 155, *156*, *158*, 159
St Stephen's Church, Bristol *21*, 22, cat4
stereotomic drawing *58*, 60
Stickells, Robert 55
Strong, Edward *164*, cat48–49
structural design methods (Wren's) 98–9, 100–4, 155, 161–4, cat45
Stuart, James 146
Stuart, John, Earl of Bute 133, 141–2, 146, 148
Sullivan, Luke *144*
surveying cross-staff 46, 62
surveying instruments 61–2, *62*, cat27
surveyor's drawings
    medieval 26, *27*, 29, cat6
    16th century *33*, 35, *38–40*, 39–44, *42–3*, cat13–14
surveyor's folding rule 50–1, *50–1*, cat17
surveyor's manuals 46
Symonds, John 55–6, *56–7*, 59–60, *59*, 61–3, cat22–23, cat25
Symons, Ralph 63–4, *64*, cat28

tables 46
Tectonicon (Digges) 46–9, *46–8*, cat15–16
template drawings 18–19, 21, 22, *58*, 60, cat1
theodolite 61–2, cat27
Thorpe, John *77–8*, 79–82, *81*, cat35, cat37
tools
    masons' *20*, 21, cat2–3

Index

see also measuring rulers
Tower of the Five Orders 74, 75–6
town plans 32–3, 33, 36, 38, 39–41, 40, cat10–12
Townshend, George 150, cat78
trace italienne 31
tracing floors 18
Treatise of Such Mathematical Instruments, As Are Usually Put into a Portable Case (Robertson) 126, 128, 128, cat60
Treatise on Civil Architecture (Chambers) 133
Trinity Library, Cambridge 98, 98
Trollap (or Trollope), Robert 54, 55
'turn-up' compasses 130, 130
Tuttell, Thomas 118, 119–20, cat53

Vertue, William 28–9, 29, cat7–8

Vignola, Giacomo Barozzi da 67, 68, cat29
Villa Pisani 77, 80
Vitruvius Pollio, Marcus 66–72
Volpe, Vincenzo 33, 35
volute compass 140–1, 142, cat73–74

water supply maps 24–5, 26
weather clock (Wren's) 85, 86–7, 87, cat38
Webb, John 83
Wells Cathedral 18
Westminster Abbey 28, 29
Wilkins, John 86
Willis, Thomas 89, 91
Winchester
    Cathedral 29, 29, cat8
    College 23, 25–6, cat5

Woodroofe, Edward 158
Worcestre (or Worcester), William 21–2, 21, cat4
working drawings 97, 105–9
Wren, Christopher
    drawing methods and skill 85–93, 86–94, 96, cat38–42
    early career 84–6, 94–6
    old St Pauls 93, 94–5, cat43
    Parentalia 98
    plans for rebuilding London 95–6, 95, cat44
    science and architecture 96–7, 109–10
    St Paul's Cathedral 97, 97, 99–100, 101, 106
    dome design 102–4, 155–69, cat45
    structural design methods 98–9, 98–9, 100–4
Wynne, Henry 115, cat51

Photographic Acknowledgements

The Warden and Fellows of All Souls College, Oxford: 73, 75, 78, 79, 80, 81, 82, 89, 143, 137
Martin Biddle: 6
Bibliothèque Nationale de France: 148
Bodleian Library, Oxford: 19, 33, 34, 35, 37, 66, 72, 85, 94, 95, 104, 107, 110, 113, 141
British Library: 4, 16, 21, 22, 23, 24, 25, 26, 27, 28, 29, 30, 31, 32
The Trustees of the British Museum: 50, 86, 100, 101, 130, 131, 133
The Carpenters' Company: 54
London Charterhouse: 13
Chatsworth Settlement Trustees: 57, 58
The Master and Fellows of Corpus Christi College, Cambridge: 9
Courtauld Institute of Art: 65
Howard Dawes: 102, 105, 106, 108, 109, 123, 124, 132
The Master and Fellows of Emmanuel College, Cambridge: 56
English Heritage (Derek Kendall): 154
Gainsborough's House: 134
Gordon Higgott: 138, 139, 140, 144, 148, 150, 155
Edward Impey: 84
The Master and Fellows of Jesus College, Cambridge: 99
London Metropolitan Archives: 15
The Master and Fellow of Magdelene College, University of Cambridge: 10
Mary Rose Trust: 36
Museum of Archaeology and Anthropology, University of Cambridge: 5
Museum of London Archaeology Service/Helen Jones: 14
Museum of the History of Science, University of Oxford: 8, 38, 39, 40, 51, 52, 63, 64, 83, 96, 97, 98, 103, 111, 114, 126, 129, 135, 136
The National Archives: 45, 46, 47, 49
National Maritime Museum: 53
National Museums Scotland: 44
National Portrait Gallery: 55
RIBA Library Drawings and Archives Collections: 17, 18, 48, 60, 125
The Royal Collection © 2009, Her Majesty Queen Elizabeth II: 115, 116, 117, 118, 119, 120, 121, 122
The Royal Society of London: 74
The Dean and Chapter of St Paul's Cathedral: 87, 88, 90, 91, 92, 93, 142, 145, 149, 151, 152, 153
Science Museum, London: 7, 43, 112, 127, 128
The Trustees of Sir John Soane's Museum: 67, 69, 70, 146, 147
The Master and Fellows of Trinity College, Cambridge: 12, 84
Wellcome Library, London: 76, 77
Whipple Museum of the History of Science, University of Cambridge: 41, 42
Winchester College: 11, 20
The Provost and Fellows of Worcester College, Oxford: 59, 61, 62, 68, 71, 157, 158